Sensors

Topical Conference at the 2010 AIChE Annual Meeting

Salt Lake City, Utah, USA
7-12 November 2010

ISBN: 978-1-61782-140-0

Printed from e-media with permission by:

Curran Associates, Inc.
57 Morehouse Lane
Red Hook, NY 12571

Some format issues inherent in the e-media version may also appear in this print version.

Printed by Curran Associates, Inc. (2011)

For permission requests, please contact AIChE
at the address below.

AIChE
3 Park Avenue
New York, NY 10016-5991

Phone: (203) 702-7660
Fax: (203) 775-5177

www.aiche.org

Additional copies of this publication are available from:

Curran Associates, Inc.
57 Morehouse Lane
Red Hook, NY 12571 USA
Phone: 845-758-0400
Fax: 845-758-2634
Email: curran@proceedings.com
Web: www.proceedings.com

TABLE OF CONTENTS

(157a) Enhanced Sensitivity of Resonating Nanocantilever Chemical Vapor Sensors Via Localized Sorptive Film Deposition

Heather C. McCaig[1], Edward B. Myers[2], Derrick Chi[2], Xinchang Zhang[2], Nathan S. Lewis[1] and Michael L. Roukes[2], (1)Chemistry, California Institute of Technology, Pasadena, CA, (2)Applied Physics, California Institute of Technology, Pasadena, CA

Sensors that are fast, small, sensitive, and can be integrated with standard microelectronics processing techniques are needed for emerging technologies such as wearable chemical vapor detectors. A promising type of sensor for these applications is the resonating cantilever, which has demonstrated detection of volatile organic compounds (VOCs), chemical warfare agents, and explosives, and has also been integrated into arrays for discrimination and identification of chemical vapors.[1] With the advent of nanoprocessing technologies, cantilever dimensions have been pushed to the nanoscale in pursuit of ever more sensitive detectors. For example, shrinking cantilever dimensions to the nanoscale greatly reduces viscous damping, allowing them to maintain high Q in ambient conditions where vapor detection occurs.[2] A high Q is required to measure small changes in the device's resonant frequency, which correspond to sorption of tiny quantities of analyte molecules. Additionally, nanocantilever sensors exhibit very high mass sensitivity, and are able to detect as little as 1 ag (10^{-18} g) of sorbed vapor.[2] However, an examination of the cantilever response mechanism shows that even further increases in the sensitivity of cantilever based sensors may be possible by selecting the response regime via device fabrication and surface functionalization.

Cantilever resonance frequency is affected by the device spring constant and effective mass, both of which change upon sorption of an analyte vapor. The relative change in resonance frequency can be approximated by:

$$\frac{\Delta f}{f_o} = \frac{1}{2}\left(\frac{\delta k}{k} - \frac{\delta m_e}{m_e}\right) \qquad (1)$$

where Δf is the change in resonance frequency, f_o is the resonance frequency, m_e is the effective mass, and k is the spring constant respectively.[3] Since sorption of analyte molecules generally causes both mass and spring constant to increase, the sum of these factors is smaller than their individual contributions to the sensor response. Eliminating sorption on the legs removes the influence of spring constant term, effectively increasing the observed response. Sensitivity can also be enhanced by allowing for vapor sorption only on the legs and relying instead solely on the spring constant response. To accomplish this, localized deposition of sorptive films is required, necessitating the use of bottom up approaches due to the size constraints of nanocantilever sensors.

Figure 1: Nanocantilever with chromium mask deposited on the tip over integrated thin gold film used for thermoelastic actuation and piezoresistive readout of the frequency signal.

Here, we demonstrate the selective deposition of sorptive films on nanocantilever sensors by utilizing the well-known self assembly of thiols onto gold surfaces.[4] To prevent film growth on either the nanocantilever tip or legs, we first deposit an additional masking layer of chromium onto the cantilever surface. We pursue two strategies to selectively deposit sorptive films on only the exposed gold regions. Our first strategy is a self assembled multilayer scheme, wherein a film is built up via sequential monolayer assembly of 11-mercaptoundecanoic acid and copper ions from copper(II) perchlorate hexahydrate in a sandwich structure.[5] Previously, this has been demonstrated as an advanced lithography technique,[6] and we can grow multilayers in excess of 30 nm thick. Our second strategy is a one-step self assembly of thiol-terminated, long-chain polyethylene glycol (PEG) onto the exposed gold surface. In both cases, ellipsometry and x-ray photoelectron spectroscopy (XPS) are used to characterize the films, and changes in nanocantilever resonance frequency are used to verify film deposition onto devices. The functionalized cantilevers are then exposed to a series of analyte vapors under dry (nitrogen background flow) and mildly humid (laboratory air background flow) conditions to determine their performance. The performance of nanocantilevers with sorptive films deposited only on the device tip or on the legs is compared to the performance of uncoated cantilevers and to that of coated cantilevers without masked regions.

This work offers the first proof of concept for the benefit of localized sorptive film deposition on nanoscale cantilever chemical vapor sensors. From here, surface functionalization can be tailored to target specific vapor analytes of interest such as nerve agents or explosives. Also, localized deposition of chemically-varied films can be pursued toward the creation of nanocantilever arrays, such as have already been demonstrated with polymer coated devices.[7] This work can also be extended to micro and nanoscale biosensors, where control over binding location of biomolecules could greatly enhance detection of very low concentration analytes.

References:

1. Zougagh, M.; Rios, A., Micro-electromechanical sensors in the analytical field. *Analyst* **2009**, 134 (7), 1274-1290.

2. Li, M.; Tang, H. X.; Roukes, M. L., Ultra-sensitive NEMS-based cantilevers for sensing, scanned probe and very high-frequency applications. *Nat. Nanotechnol.* **2007**, 2 (2), 114-120.

3. Chen, G. Y.; Thundat, T.; Wachter, E. A.; Warmack, R. J., Adsorption-induced surface stress and its effects on resonance frequency of microcantilevers. *Journal of Applied Physics* **1995**, 77 (8), 3618-3622.

4. Bain, C. D.; Troughton, E. B.; Tao, Y. T.; Evall, J.; Whitesides, G. M.; Nuzzo, R. G., FORMATION OF MONOLAYER FILMS BY THE SPONTANEOUS ASSEMBLY OF ORGANIC THIOLS FROM SOLUTION ONTO GOLD. *J. Am. Chem. Soc.* **1989**, 111 (1), 321-335.

5. Evans, S. D.; Ulman, A.; Goppertberarducci, K. E.; Gerenser, L. J., SELF-ASSEMBLED MULTILAYERS OF OMEGA-MERCAPTOALKANOIC ACIDS - SELECTIVE IONIC INTERACTIONS. *J. Am. Chem. Soc.* **1991**, 113 (15), 5866-5868.

6. Srinivasan, C.; Hohman, J. N.; Anderson, M. E.; Weiss, P. S.; Horn, M. W., Nanostructures using self-assembled multilayers as molecular rulers and etch resists. *J. Vac. Sci. Technol.* B **2007**, 25 (6), 1985-1988.

7. McCaig, H. C., Myers, E. B.; Chi, D.; Zhang, X.; Lewis, N. S.; Roukes, M. L., Polymer-coated nanocantilever sensor array for discrimination of volatile organic compounds (VOCs). In *American Chemical Society Spring Meeting*, 2010.

See more of this Session: Micro and Nanofabricated Sensors
See more of this Group/Topical: Topical 9: Sensors

(157b) On Chip Electrochemical Detection of Biomarkers for Detection of Water Borne Toxins

Rebekah C. Wilson, U.S. Army ERDC - CERL, Champaign, IL, Donald M. Cropek, Construction Engineering Research Laboratory, US Army Engineer Research and Development Center, Champaign, IL and Scott A. Banta, Chemical Engineering, Columbia University, New York, NY

Clean water is not a luxury but a necessity. While clean water is bountiful in the United States, the same cannot be said for developing countries. It would be advantageous to not only monitor water quality without the constraints of a laboratory but to be able to pinpoint specific organs that might be at risk. The devices presented offer such a solution by combining microfluidics with electrochemical detection to monitor real time response of organs to water borne toxins. This research builds on recent scientific advances in microfabrication of human organ structures to further create new simulated organs containing the relevant chemical functionality that mimics the toxic responses as well as detection schemes to measure assault of toxins in specific organs. Organs of interest (heart, liver, etc) are mounted within separate parallel channels with a glass/PDMS device. Water samples are injected into the device and partitioned between the different organ cultures. Protein modified (Cytochrome C) electrodes positioned beneath each channel are then used to monitor the release of superoxide (a biomarker for toxin assault). By supplying a specific potential, the concentration of superoxide present can be monitored amperometrically. Advances towards this device will be presented including response, calibration and figures of merit.

See more of this Session: Micro and Nanofabricated Sensors
See more of this Group/Topical: Topical 9: Sensors

(157c) Electrochemical Detection in Nanofluidic Channels

Edgar D. Goluch, Nuchapong Wongrajit, Pradyumna S. Singh and Serge G. Lemay, Kavli Institute of Nanoscience, Department of Bionanoscience, Delft University of Technology, 2628 CJ Delft, Netherlands

As experiments involving micro- and nanofluidic devices become commonplace, the need for detection at these scales becomes increasingly important. Many current microfluidic sensing techniques employ optical methods. While they can be incredibly sensitive, optical techniques are often complicated and difficult to incorporate with the rest of the microfluidic device fabrication process. A promising alternative to optical sensors involves the use of electrical elements incorporated into fluidic systems. Electrical sensors can be less demanding in space, less expensive, and easier to integrate than their optical counterparts.

We have developed a nanofluidic detection scheme that specifically exploits these small dimensions. The device consists of two parallel electrodes in a nanofluidic channel with openings at either end that allows for very sensitive detection of electrochemically active species. Electrochemically active molecules that enter the cavity transfer charge between the electrodes when a potential difference is applied. Since these molecules are capable of repetitively undergoing oxidation and reduction, each molecule can transfer, on average, thousands of electrons by repeatedly traveling between the electrodes before escaping back out into the bulk.

The fabrication process involves a combination of lithographic patterning and evaporation to form the electrodes and nanocavity. We have shown the benefits of such an approach for measuring molecular fluctuations [1], absorption [2] and electron-transfer kinetics [3]. We have also demonstrated the utility of such an arrangement for sensor applications, where catechol [4] and acetaminophen [5] were detected in the presence of ascorbic acid. This presentation will focus on the fabrication, design rules, and ongoing efforts to integrate these sensors in complex microfluidic systems.

REFERENCES:

[1] M.A.G. Zevenbergen, D. Krapf, M.R. Zuiddam, S.G. Lemay, Nano Letters, 8, 497 (2007).

[2] M. A. G. Zevenbergen, P.S. Singh, E.D. Goluch, B.L. Wolfrum, S.G. Lemay, Analytical Chemistry, 81, 8203 (2009).

[3] M. A. G. Zevenbergen, B.L. Wolfrum, E.D. Goluch, P.S. Singh, S.G. Lemay, Journal of the American Chemical Society, 394, 447 (2009).

[4] B.L. Wolfrum, M.A.G. Zevenbergen, S.G. Lemay, Analytical Chemistry, 36, 3342 (2008).

[5] E.D. Goluch, B.L. Wolfrum, P.S. Singh, M.A.G. Zevenbergen, S.G. Lemay, Analytical and Bioanalytical Chemistry, 22, 1036 (2009).

See more of this Session: Micro and Nanofabricated Sensors
See more of this Group/Topical: Topical 9: Sensors

(157d) Nanowire Sensor Array Based On Mixed Metal Oxide

Xiaopeng Li[1], Junghwan Cho[2], Qingwen Yan[1], Hongwei Sun[3], Pradeep U. Kurup[2] and Zhiyong Gu[1], (1)Chemical Engineering, University of Massachusetts Lowell, Lowell, MA, (2)Civil and Environmental Engineering, University of Massachusetts Lowell, Lowell, MA, (3)Mechanical Engineering, University of Massachusetts Lowell, Lowell, MA

Nanowire has been proved to be a promising candidate for sensor applications due to its high aspect ratio, improved specific surface area and enhanced sensitivity toward certain analytes. However, most of the nanowire sensors developed so far reply on single-elements and thus work individually; hence, they face challenges in selectivity and are incapable of discrimination and classification. The introduction of a secondary element into the sensor elements has the promise to solve the problems above. Here, we present our recent research work of fabricating mixed metal oxide nanowires and further integration of a nanowire-based sensor array. Mixed metallic nanowires were first synthesized with various element combinations, including tin, indium, nickel, copper, etc. Then the nanowires were integrated on to inter-digitated microelectrodes as a sensor chip using dielectrophoretic (DEP) assembly technique. A following thermal oxidation process converted mixed metallic nanowires into their corresponding oxides. Morphology of the nanowires before and after oxidation was characterized by scanning electron microscope (SEM) along with Energy Dispersive x-ray Spectroscopy (EDS) for compositional information, while their electrical properties were examined by a sourcemeter after each step. A sensor device containing up to four sensors and an auto-sampling system were designed, respectively, and were utilized for sensor testing. From the results, enhanced sensor performance was observed for mixed metal oxide nanowire sensors with particular compositions. Also, both principle component analysis (PCA) and artificial neural network (ANN) were applied to analyze the classification ability of the sensor array.

See more of this Session: Micro and Nanofabricated Sensors
See more of this Group/Topical: Topical 9: Sensors

(157e) Implantable, Near-Real-Time Microelectrode Array Biosensor for Simultaneous Monitoring of Glutamate and Dopamine

Tina T.-C. Tseng and Harold G. Monbouquette, Department of Chemical and Biomolecular Engineering, University of California, Los Angeles, Los Angeles, CA

A Nafion and over-oxidized polypyrrole (OPPy)-modified, silicon-based platinum (Pt) microelectrode array biosensor was developed with optimal spatial and temporal resolution for simultaneous monitoring of glutamate (Glut) and dopamine (DA) by constant potential amperometry. A 2×2 microelectrode array located at the tip of a silicon-based microprobe with 100 µm and 40 µm separation between the microelectrodes on the same column (top and bottom sites) and the microelectrodes on the same row (neighboring sites), respectively, was patterned for maximum spatial resolution and minimum diffusional interaction of analytes. For each microelectrode, the average area was ~5000 μm^2. The microprobe shaft, measuring 120 µm in width and 150 µm in thickness, was designed to inflict minimum tissue damage during implantation while still providing sufficient mechanical stability. Microelectrodes modified with polymeric films provide a high degree of selectivity in the presence of electroactive interferents. Thin films of OPPy have been shown widely to have permselectivity against anions and excellent selectivity towards cations. Based on our previously reported results, however, a thick film of OPPy rejects both ascorbic acid (AA) and positively charged DA effectively. The total charge was controlled during the electrodeposition of polypyrrle (PPy) for an optimum film thickness. Selective deposition of thin PPy films at the top sites and thick PPy films at the bottom sites were followed by PPy over-oxidation. Nafion was dip-coated on all OPPy-modified microelectrodes to enhance the discrimination against negatively charged species, such as AA and uric acid (UA). A glutaraldehyde cross-linked glutamate oxidase (GlutOx) and bovine serum albumin (BSA) matrix was immobilized manually on the bottom two sites (with thick OPPy films). GlutOx turns over the non-electroactive analyte, Glut, in the presence of oxygen to give the electrooxidizable product, hydrogen peroxide, which undergoes a two-electron transfer reaction at the Pt surface at sufficiently oxidizing potential. The final configuration of the dual Glut/DA sensor was Pt/thick OPPy/Nafion/GlutOx for the Glut microsensor sites at the bottom and Pt/thin OPPy/Nafion for the DA microsensor sites at the top. Good selectivity, detection limit, sensitivity, and temporal resolution of the dual Glut/DA sensor were achieved *in vitro*. A constant oxidation potential was maintained at 0.7 V vs. Ag/AgCl. There was no detectable response from AA over the physiological concentration range. In the presence of both Glut and DA, the Glut microsensor sites responded to Glut with a detection limit of 5 µM and a 1 second response time with no DA response; whereas the DA microsensors detected DA with a detection limit of 100 nM and a 1 second response time without interference from Glut. The sensitivity of the dual sensor was ~50 $nA{\bullet}\mu M^{-1}{\bullet}cm^{-2}$ and ~1280 $nA{\bullet}\mu M^{-1}{\bullet}cm^{-2}$ toward Glu and DA, respectively. Studies of Glut and DA changes in the brains of live rodents using this microelectrode array biosensor currently are underway.

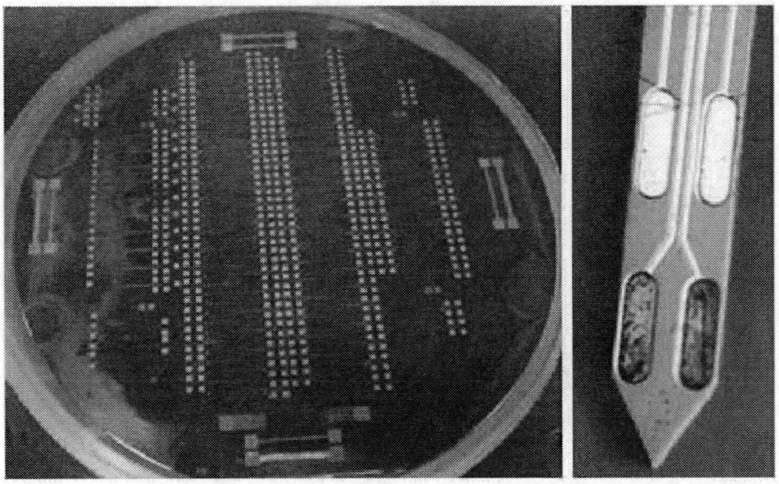

See more of this Session: Micro and Nanofabricated Sensors
See more of this Group/Topical: Topical 9: Sensors

(157f) Combination of Electrospinning and Hydrogel Patterning to Create Microarray System for Dual Sensing

Yeol Lee, Hyun Jong Lee, Kyung Jin Son, Sangphil Park, Seung Hee Nam, Young Ha Park and Won-Gun Koh, Chemical and Biomolecular Engineering, Yonsei Univ., Seoul, South Korea

Micropatterning is the important technique in developing miniaturized biosensors and most of micropatterns have been created on various planar substrates. However, the amount of biomolecules that can be immobilized onto a planar surface is limited, affecting sensor performance of relatively weak signal. In order to fully realize potential of protein biochips, proteins need to be immobilized onto surfaces with high density. Among many materials, electrospun nanofibers can provide a large surface area for the attachment of proteins, which can lead to high volumetric protein activity. In this study, we fabricate polymeric nanofibers with a mixture of polystyrene and poly(styrene-alt-maleic anhydride) (PSMA). Proteins can be immobilized onto surfaces of nanofibers without any special modification due to acid anhydride group. And also micropatterned surface was created by poly (ethylene glycol) (PEG) hydrogel microstructures via photolithography on PSMA nanofiber mats. As a result, two different regions are combinatorially fabricated in one substrate. Two different kinds of biomolecules can be detected in each region. Immunoassay system could be prepared by immobilizing mouse immunoglobulin G (IgG) on PSMA nanofibers and different concentration of fluorescence labeled rabbit anti-mouse IgG was used. Glucose oxidase (GOX) and Peroxidase are entrapped in hydrogel region. The GOX-catalyzed oxidation reaction was investigated with Amplex Red and different concentrations of glucose.

See more of this Session: Micro and Nanofabricated Sensors
See more of this Group/Topical: Topical 9: Sensors

(157g) A Microfabricated Stretchable Antenna for Wireless Strain Sensing

MIchael D. Dickey[1], Ju-Hee So[1], Amit Qusba[2], Gerard J. Hayes[2] and Gianluca Lazzi[2], (1)Chemical and Biomolecular Engineering, North Carolina State University, Raleigh, NC, (2)Electrical and Computer Engineering, North Carolina State University, Raleigh, NC

This talk will describe a new microfabrication strategy that uses a moldable liquid metal to produce highly stretchable fluidic antennas that serve as simple wireless sensors of strain. Conventional electronics are typically fabricated from rigid materials (e.g., silicon for transistors, copper for antennas). New materials are being explored as candidates for flexible electronics because of the novel applications that emerge from their mechanical properties. Examples include flexible displays, implantable devices, and electronic textiles. While the flexibility of current devices stems from thin-film geometries, rigid materials cannot be stretched or deformed significantly without inducing irreversible damage. Liquids flow in response to stress and are therefore highly deformable; most liquids, however, have low electrical conductivities, tend to evaporate, and cannot be micromolded into desirable shapes because they typically resolve to hemispherical shapes to minimize surface energy. Injecting a gallium-based metal alloy into elastomeric microchannels offers a new method of microfabricating flexible electronics. The metal is a liquid at room-temperature with low-viscosity (water-like) and has mechanical properties that are governed by a thin, oxide skin that forms rapidly on its surface in the presence of oxygen. The presence of the skin allows the low-viscosity fluid to form mechanically stable components in microfluidic channels at room temperature. Since the metal flows in response to stress, the mechanical properties of the fluidic antennas are defined by the elastomeric encasing material. The talk will discuss the properties of the metal and methods to mold the metal to form ultra-stretchable sensors.

See more of this Session: Micro and Nanofabricated Sensors
See more of this Group/Topical: Topical 9: Sensors

Liquid Crystal Droplets as a Hosting and Sensing Platform for Developing Immunoassays

Vera Joanne Aliño, Jasmine Pang and Kun-Lin Yang*
Department of Chemical and Biomolecular Engineering, National University of Singapore, Singapore

Summary

Protein immunoassays generally consist of the immobilization of the proteins on solid surfaces followed by binding with their antibodies. Such systems often require several washing and processing steps which are both time consuming, costly and labor intensive. In this study, we developed an immunoassay in which the proteins are directly immobilized on the surface of the liquid crystal (LC) droplets rather than on solid surfaces. The advantage of this assay is that the binding of antibodies to the surface-immobilized proteins can be transduced by the LC droplets without the need for additional steps for developing readout signals. For example, we observed that after incubation of IgG-decorated LC droplets in an aqueous solution containing 50µg/mL of anti-IgG (AIgG) for 15 min, the orientation of the LCs changed from a radial to a bipolar configuration. In contrast, the orientation of the LCs remains unchanged after the incubation in an aqueous solution containing anti-HSA (AHSA) of the same concentration. We conclude that the bipolar configuration of LCs is a signature of the presence of specific antigen-antibody pairs on the surface. By using this immunoassay, we successfully detect protein concentrations as low as 0.005µg/mL for IgG and 0.01µg/mL for HSA. This is more sensitive than current immunoassays on solid surfaces. In addition, the immunoassay has a dynamic range of 10^4. These results demonstrate the potential of using LC droplets as sensing platforms for the development of rapid and low-cost immunoassays.

Keywords: *protein immunoassays, protein detection, liquid crystals, LC droplets*

Extended Abstract

Over the past decade, liquid crystals (LCs) have shown great potential in the development of label-free platforms for detecting the presence of chemical and biological agents [1]. The detection principle is based on the highly sensitive response of orientational ordering of LC phases to minute changes on surfaces on which LCs are supported. The LCs at the surface can then communicate their orientations through the bulk, causing global changes in orientations and generating optical textures that can be readily observed with the naked eye under cross polarizers. Recently, this mechanism has been exploited to detect the presence of proteins on solid surfaces or at aqueous/LC interfaces. For example, our group demonstrated a LC-based protein immunoassay which allows the detection and quantification of different antibodies that bind to antigens immobilized on solid surfaces [2]. This LC-based protein immunoassay offers advantages over conventional immunoassays because it does not require complex instrumentation. However, similar to other existing immunoassays, it still requires several steps before the test results can be obtained. These steps include surface treatment of the glass slides, antigen and antibody immobilization, rinsing in between steps, and lastly, the detection analysis including the fabrication of the LC cell and filling the cell with LCs.

In the past, researchers have shown interest in making LCs in droplet forms because of their high potential as materials for various electrooptical devices such as optical switches and light valves. Recently, it has been proposed that the spherical configuration of the LC droplets may have greater potential for sensing applications since they have larger surface areas than the LC films at solid/planar interfaces. Furthermore, LC droplets are mobile and their homogenous distribution within the solution can lead to faster diffusion and shorter response time. Several methods for the fabrication and characterization of LC droplets have been reported. Tongcher, *et. al.* for instance, produced stable LCs droplets in aqueous systems by using hydrophobe such as hexadecane to suppress molecular diffusion degradation (Ostwald ripening) and coalescence by collision [3]. In addition, they also reported that ionic and steric surfactants such as SDS and SE3030 can improve the stability of the LC droplets. Moreover, LC droplets can also be stabilized in aqueous solution using water-soluble polymers according to Simon and co-workers [4]. Sivakumar, *et. al.* created monodispersed LC droplets starting from silica particles coated with multilayer of polyelectrolytes on the outside. The silica particles were etched with hydrofluoric acid (HF), allowing the hollow polyelectrolyte capsule to be filled with LCs. This method, however, has its disadvantages as it requires a clean room technology which can be time consuming and labor intensive. Hence, using such as system limits the freedom to use it for further applications [5].

In this work, we aimed to use LC droplets as a platform for hosting protein immunoassays. So far, there is still no report of using LC droplets for hosting immunoassays. We also studied the immobilization of proteins on the surface of LC droplets and investigated how the presence of protein and immunocomplex affect the orientations of LC inside the droplets. We investigate the production of LC droplets using polyethylene imine (PEI) to encapsulate the droplets as well as control the droplet size. We focused on detecting two urinary proteins particularly IgG (human immunoglobulin G) and HSA (human serum albumin) which are important biomarkers for chronic kidney disease.

Materials and methods

LC droplets were first prepared from 5CB mixed with 6%w/w hexadecane as hydrophobe. The 5CB-hexadecane mixture was then added to aqueous PEI solution. PEI concentration was studied to find optimum conditions. IgG and HSA were then immobilized to the LC droplets using 0.5 wt% glutaraldehyde (GA) as linker. Excess aldehyde groups were passivated using 0.1 mM ethanolamine. Protein-decorated LC droplets were then immersed in Tween 20 buffer solution (PBS-T20) containing 50µg/mL Cy3-AIgG or Cy3-AHSA at room temperature. Bright-field, crossed polars, and fluorescence images of the LC droplets were acquired using an optical microscope (Nikon ECLIPSE LV100POL, Japan) equipped with a digital camera. Fluorescence images were captured using Nikon HYQ fluorescence filter set. To determine the droplet size, three images were taken at different locations of the optical cell and the droplets were analyzed using the Nikon's imaging software (NIS-Elements).

Results and discussions

We first studied how to control the surface density of PEI on the LC droplets by varying the PEI concentration and then studied the effects of PEI on the droplet size and its stability. Figure 1 shows that the surface density of PEI on the LC droplets increases proportionally to the PEI concentration. It should be noted that having a higher PEI surface density is favored because it gives more reactive amine groups to immobilize the proteins.

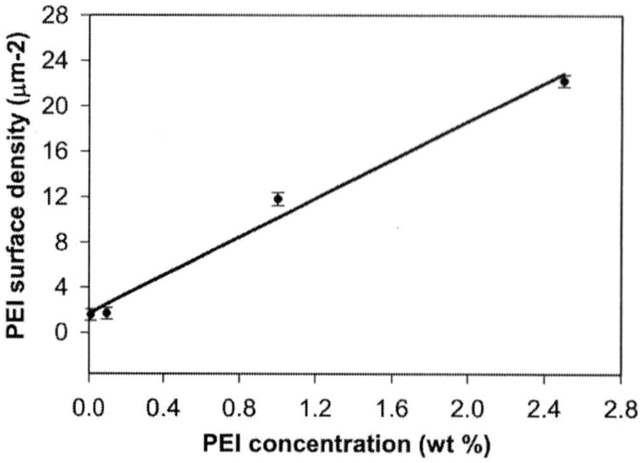

Figure 1. Effect of PEI concentration on the surface density of PEI on the LC droplets.

Next, we studied the effect of PEI concentration on the size of the LC droplets. Table 1 shows a decrease in the droplet size with increasing PEI concentration. This is probably caused by the increase in the viscosity of the continuous phase. For a given vortexing speed, shear stress is stronger when the viscosity of the continuous phase is higher; hence, smaller droplets are produced in the mixer. In addition, increasing the PEI concentration also increases the number PEI molecule that covers the surface of LC droplets, and that prevents the coalescence of smaller LC droplets. Although there is a decreasing trend in the droplet size, the size of the LC droplet only decreases by 8% when the PEI concentration is increased from 0.1% to 1.0%. Therefore, we hypothesize that if we further increase the PEI concentration above 1.0%, then the decrease in the droplet size is negligible.

Aside from having higher amine groups on the surface, it is also important to obtain LC droplets with good stability since we are using the droplets for developing immunoassays. Therefore, we monitored the size of LC droplet as a function of storage time. As shown in Table 1, the sizes of LC droplets prepared from 0.01% and 0.1% of PEI solution gradually increases over a period of 20 days, indicating that LC droplets prepared by using these two solutions are not stable. However, the size of LC droplets prepared from 1.0% PEI solution only increases by 14% (74 ± 28µm) after 20 days. Because 1.0% PEI solution provides higher amine surface density and better stability for LC droplets, we used 1% PEI solution to prepare LC droplets in our subsequent experiments.

Table 1. Effect of PEI concentration on the LC droplet size (µm) and droplet stability over a period of 20 days

Days	PEI concentration (wt %)		
	0.01%	**0.10%**	**1.00%**
2	115 ± 51	71 ± 33	65 ± 18
6	131 ± 71	93 ± 41	67 ± 19
14	225 ± 223	106 ± 28	65 ± 10
20	308 ± 156	157 ± 50	74 ± 28

We studied the optical textures of the LC droplets under a bright-field microscope. We observed two points of defects (boojums) found at the edge of the droplets as indicated by the white arrows (Figure 2A). We also investigated the birefringence properties of the LCs by viewing them under cross polarizers. The polarized-light image displays bright spherical dots with dark dimples where the boojums are located. Both images suggest that the orientations of LCs inside the droplets exhibit bipolar configuration. These results are consistent with other reports on LC droplets with sizes > 3μm [6]. Systems with bipolar configuration occur when LC molecules orient parallel or tilted to the surface.

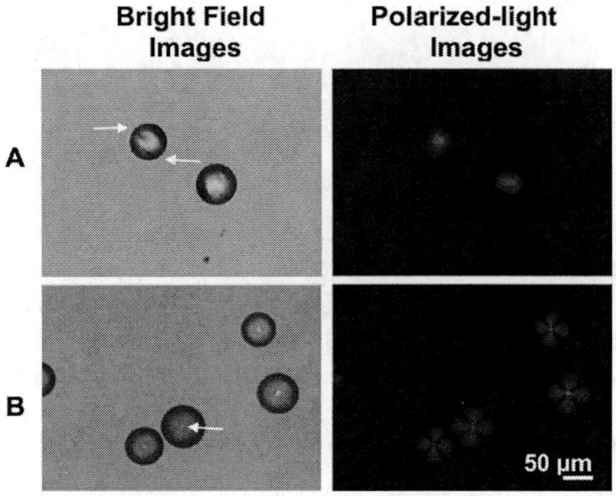

Figure 2. Images of the LC droplets before (A) and after (B) immersing in PBS-T20 buffer causing the LC droplets to exhibit a bipolar and radial configuration, respectively.

On the other hand, when we dispersed the LC droplets in PBS-T20 buffer, only a single boojum located at the center of the droplet was observed as shown in Figure 2B. Under cross polarizers, each bright spherical dot has a dark cross in it. Therefore, we conclude that orientations of LCs in these droplets are in radial configuration. Such configuration occurs when the LC molecules at the interface orients perpendicular to the surface of the droplets. We interpret from the above results that the Tween 20 molecules can permeate through the polymer film and because of their amphiphilic structure, Tween 20 can position at the interface between the bulk solution and the LCs, thus inducing the orientation of the LCs in the droplet. Lockwood, *et. al.* also observed similar phenomenon when they exposed a thin film of LCs in an aqueous solution to a linear surfactant, sodium dodecyl sulfate (SDS) [7]. They found that the polymer PSS-PAH bilayers at the interface can mediate the interactions between the surfactant added to the aqueous phase and the surface of the LCs. A schematic illustration of the orientation of LCs in the droplet before and after the addition of Tween 20 is shown in Figure 3.

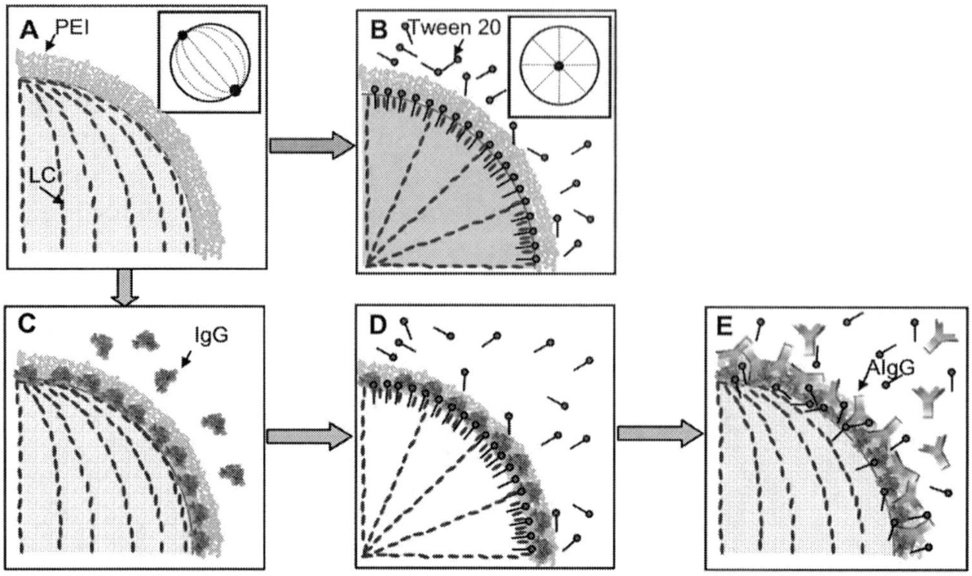

Figure 3. Schematic illustration of the LC droplet during chemical functionalization and protein immobilization. (A) A LC droplet in PBS buffer exhibits a bipolar configuration. (B) Addition of Tween 20 in PBS buffer (PBS-T20) induces an orientational transition from a bipolar to a radial configuration. (C) Immobilization of IgG on the surface of the LC droplet. (D) Rinsing the IgG-LC droplets with PBS-T20 buffer exhibits a radial configuration suggesting that the T20 molecules can still diffuse to the interface even if low density proteins are present on the surface. (E) Immunocomplex formation on the surface changes the LC orientation to a bipolar configuration.

In order to immobilize IgG at the surface of the LC droplet, GA was reacted with PEI to provide free aldehyde groups for protein immobilization. We determined the effect of the amount of GA on the orientation of the LCs in the droplet. Figures 4A and 4B show that at 0.1% and 0.5% GA, the LC droplets still exhibit a radial configuration. However, at 1% GA the orientation has already changed to a bipolar configuration (Figure 4C). This result confirms that the orientation of LCs at the interface is also affected by the chemical functionality and the amount of molecules attached to it. Since a radial configuration is preferred, the GA concentration was fixed at 0.5% in the following experiments.

Figure 4. Effect of GA concentration on the LC orientations of the droplets. Polarized- light images of the LC droplets when GA concentrations 0.10% (A), 0.50% (B) and 1.00% are used.

Next, we immobilized IgG on the surface of the GA-functionalized LC droplets. As shown in Figure 5A, the IgG-LC droplets exhibit a radial configuration. Thus, we conclude that the presence of IgG does not affect the orientation of LCs in the droplet.

We used two fluorescently labeled antibodies Cy3-AIgG and Cy3-AHSA to study its immunobinding with the IgG-LC droplets. Figures 5B and 5C shows the fluorescence and polarized-light images of the IgG-LC droplets after the reaction with Cy3-AIgG and Cy3-AHSA, respectively. The red fluorescence dots indicate that AIgG binds to the IgG immobilized on the surface of the LC droplet. On the other hand, a dark fluorescence signal after the reaction with Cy3-AHSA suggests the absence of nonspecific binding. Next, we compared the optical textures of the IgG-LC droplets before and after the immunoreaction with the antibodies. The polarized-light images indicate that the IgG-LC droplets change to the bipolar configuration after the reaction with Cy3-AIgG. We propose that this change in the orientation of LCs (from radial to bipolar) is caused by the formation of immunocomplex when AIgG binds to the surface of the IgG-LC droplets. The presence of the immunocomplex on the surface hinders the diffusion of the Tween 20 molecules at the interface. Moreover, the immunocomplex can cause a disruption in the perpendicular (homeotropic) LC orientations at the interface and thus changes the orientational configuration in the LC droplets. In contrast, the orientations of the IgG-LC droplet retain its radial configuration after the addition Cy3-AHSA. This result further confirms that the change is due to the formation of immunocomplex on the surface of LC droplets.

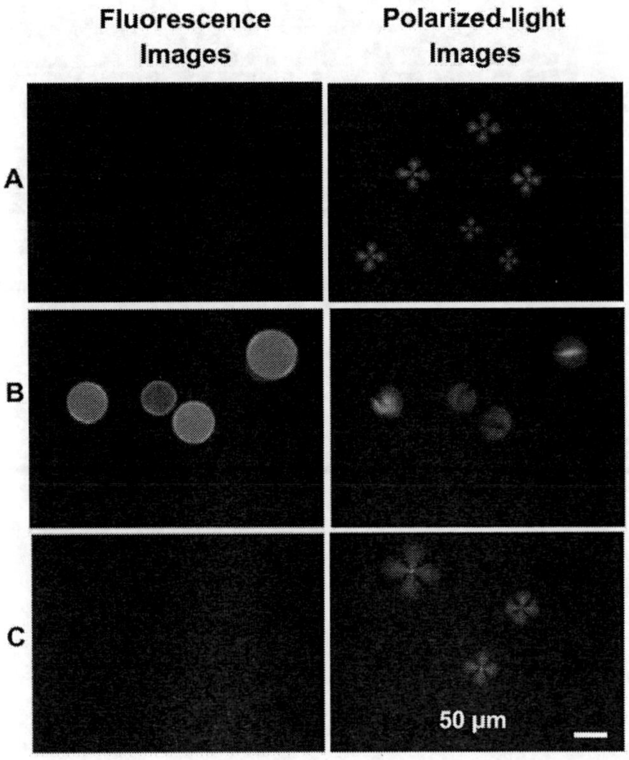

Figure 5. Images of IgG-LC droplets before (A) and after immersing with AIgG (B) and AHSA (C) solution.

To investigate whether this LC-based immunoassay can be applied to detect other proteins, we repeated the experiment by using HSA-LC droplets. Fluorescence images before (Figure 6A) and after (Figure 6B) the immersion of HSA-LC droplets in PBS-T20 buffer containing Cy3-IgG shows that there was no reaction as shown by the negative fluorescence image and that the radial configuration still remains. In contrast, when HSA-LC droplets were immersed in a Cy3-AHSA solution (Figure 6C), red fluorescence spots were obtained indicating that AHSA binds to the HSA on the surface. This is accompanied by the change of orientation of the LCs from radial to bipolar configuration.

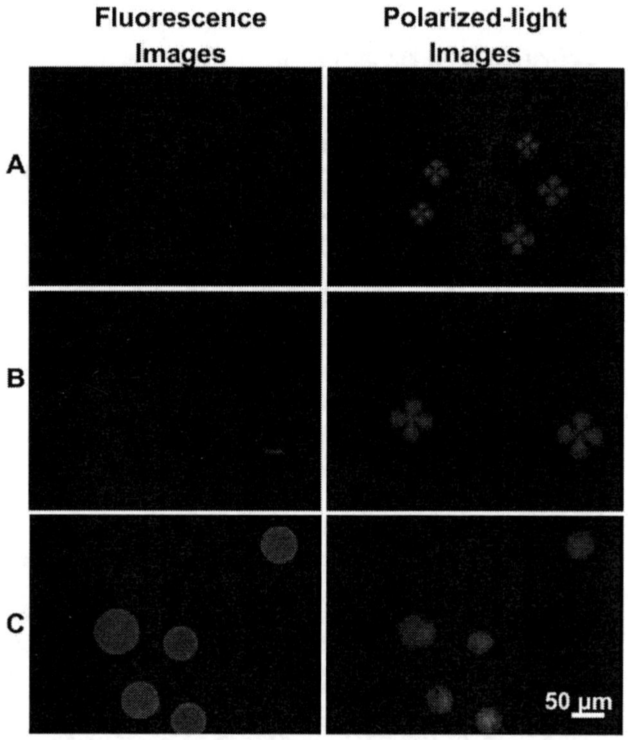

Figure 6. Images of HSA-LC droplets before (A) and after immersing with AIgG (B) and AHSA (C) solution.

Conclusions

We have demonstrated that LC droplets can be used as a hosting and sensing platform for developing protein immunoassays. We also demonstrated that the LC droplets remained stable at 1 wt% of PEI. The presence of the specific protein can be directly monitored by the optical texture produced by the LC droplet. This sensing mechanism is based on the formation of immunocomplexes between antibody and antigen immobilized on the surface of LC droplets. The immunocomplex is able to disrupt the homeotropic ordering at the interface of the LC droplets and causes a bipolar configuration and bright optical readout signals.

References

1. Brake, J. M., Mezera, A. D., and Abbott, N. L. *Langmuir* **2003,** *19*, 6436-6442.
2. Xue, C.-Y., Khan, S. A., and Yang, K.-L. *Advanced Materials* **2008,** *20*, 1-5.
3. Tongcher, O., Sigel, R., and Landfester, K. *Langmuir* **2006,** *22*, 4504-4511.
4. Simon, K. A., Sejwal, P., Gerecht, R. B., and Yang-Yeung, L. *Langmuir* **2007,** *23*, 1453-1458.
5. Sivakumar, S., Gupta, J. K., Abbott, N. L., and Caruso, F. *Chemistry of Materials* **2008,** *20*, 2063-2065.
6. Tjipto, E., Cadwell, K. D., Quinn, J. F., Johnston, A. P. R., Abbott, N. L., and Caruso, F. *Nano Letters* **2006,** *6*(10), 2243-2248.
7. Lockwood, N. A., Gupta, J. K., and Abbott, N. L. *Surface Science Reports* **2008,** *63*, 255-293.

(203b) Bioconjugation of Optical Microcavities for Label-Free Sensing

Heather K. Hunt, Mork Family Department of Chemical Engineering and Materials Science, University of Southern California, Los Angeles, CA, Carol Soteropulos, Department of Biomedical Engineering, University of Southern California, Los Angeles, CA and Andrea M. Armani, Mork Family Department of Chemical Engineering and Materials Science & Ming Hsieh Department of Electrical Engineering-Electrophysics, University of Southern California, Los Angeles, CA

The development of biosensors with high sensitivity and specificity is of significant interest to the fields of medical diagnostics and environmental monitoring, where rapid and real-time detection of antigens, bacteria, viruses, etc., is necessary. Traditional, labeled sensors, such as fluorescent immunoassays, are not capable of real-time detection, and detect the presence of the fluorescent probe rather than the molecule of interest itself. In the past decade, a new class of sensors consisting of high-performance optical devices has been developed based on whispering gallery microcavities. Initially designed for telecommunications applications, these structures have demonstrated unique capabilities in the biosensing arena primarily due to their very low optical loss. The low optical or material loss of these devices enables "optical amplification" of otherwise undetectable signals to occur within the structure, improving the signal to noise ratio. This optical loss metric is quantitatively characterized by the quality factor (Q) of the cavity. The two most commonly used microcavities, the microsphere and the microtoroid, have quality factors in excess of 100 million, which corresponds to photon storage times of greater than 100 ns. As a result of their circular geometry, the optical field orbits at the periphery of the device and is not completely confined within the cavity. This creates an evanescent tail which interacts with the environment, enabling detection of chemical or biological species. For example, using silica microcavity devices, researchers have shown label-free detection of protein conformation changes and bacteria, as well as label-free detection of virus and molecules at the single molecule concentration level.[1-2]

Previous research efforts have focused attention on the development of the sensor itself. However, while the sensitivity of the device is dependent on the low optical loss, specificity is an equally if not more important feature of any sensing platform. Therefore, it is crucial to develop a high density, covalent surface functionalization process which also maintains the optical device's performance metrics. In the case of the whispering gallery mode sensor, the most important parameter is the Q factor of the cavity. Here, we demonstrate a facile method to impart specificity to optical microcavities without adversely impacting their optical performance. We have used the silica ultra-high-Q microtoroid microcavity as the test platform, because it is the only microcavity fabricated on a silicon substrate which has achieved Q factors in excess of 100 million.[3] However, the techniques developed are transferable to other optical cavities, such as microrings, microspheres and microcylinders.

In this approach, we selectively functionalize the surface of the silica microtoroids with amineterminated silane coupling agents of various lengths using two different methods: 1) organic solvent deposition and 2) vapor deposition grafting techniques. The surface is then biotinylated via N-hydroxysuccinimide (NHS) ester chemistry at the amine end of the linker under typical literature conditions. The as-fabricated and surface-modified devices are characterized qualitatively by optical microscopy and scanning electron microscopy as well as quantitatively by microcavity analysis techniques (Q factor) to determine the impact of the surface functionalization methods on the device sensitivity, and to evaluate the conditions best suited to ensuring the devices' performance. Additionally, the surface chemistry and properties of these devices are explored via X-ray photoelectron spectroscopy, contact angle measurements, and fluorescent imaging at each reaction step, and show high density surface coverage of only the devices themselves. Based on these analyses, the optimal parameters for fabrication and surface modification protocols have been found, and result in the creation of devices whose surface functionalities have high density packing, low optical absorption, specificity only to the target ligand, and high stability in air and water, as well as high device sensitivity. Additionally, the resulting biotinylated device is used to detect streptavidin as a proof of concept.

This work represents one of the first examples non-physisorption-based bioconjugation of optical microtoroid resonators for the label-free detection of biomolecules. The primary advantage of a device that has been functionalized using the methods described above is that a wide array of biomolecules can be detected and studied, since the functionalization process does not limit the study to a specific ligand-substrate pair, or even a specific type of molecule. For example, the protocols developed here are easily extended to a wide variety of probe molecule / target molecule pairings by using the high affinity of biotin for (strept)avidin-modified probes molecules, allowing the easy interchange of the terminal functional group probes on the surface of the devices. Additionally, these methods can also be extended to a variety of silica-based optical sensors, such as microdisk and microsphere resonators, as well as waveguide-based sensors, resulting in an array of high specific sensor devices that can be utilized for rapid and real-time detection. Lastly, the optical biosensors developed during this work may be used to probe macromolecular interactions at the single-molecule level due to their ultra-high sensitivity, possibly providing insight into rare conformations, unusual reaction pathways, and fluctuating systems. This could lead to a detailed understanding of interactions that would be impossible with ensemble studies, where the behavior of individual molecules is easily disguised within ensemble averages.[4]

1. Armani, A. M.; Kulkarni, R. P.; Fraser, S. E.; Flagan, R. C.; Vahala, K. J. Science 2007, 317, (5839), 783-787. 2. Vollmer, F.; Arnold, S. Nature Methods 2008, 5, (7), 591-596. 3. Armani, D. K.; Kippenberg, T. J.; Spillane, S. M.; Vahala, K. J. Nature 2003, 421, (6926), 925-928. 4. Sakamoto, T.; Webb, M. R.; Forgacs, E.; White, H. D.; Sellers, J. R. Nature 2008, 455, (7209), 128-U99.

See more of this Session: Biosensor Devices
See more of this Group/Topical: Topical 9: Sensors

(203c) Dual Enzyme Biosensor for Detection of Organophosphorus Compounds Using Organophosphorus Hydrolase and Horseradish Peroxidase

Asli Sahin[1], Scott Banta[1], Alan C West[1] and Donald M. Cropek[2], (1)Chemical Engineering, Columbia University, New York, NY, (2)Construction Engineering Research Laboratory, US Army Engineer Research and Development Center, Champaign, IL

Organophosphorus (OP) compounds are neurotoxic compounds that are commonly used as chemical warfare agents and pesticides. Due to their high toxicity, rapid and sensitive field detection of these compounds has been a topic of interest. Previous work in this field involving organophosphorus hydrolase (OPH) has focused on direct amperometric detection of the electroactive hydrolysis products and potentiometric detection of local pH changes. We have developed a novel amperometric dual enzyme biosensor for detection of a large class of OP compounds. The working principle of this biosensor is hydrolysis of OP compounds by organophosphorus hydrolase followed by electrochemical detection of the phenolic leaving groups by horseradish peroxidase. Detection of the phenolic leaving group is carried out at -50mV vs. Ag/AgCl where the noise and background are low and interferences are negligible. Optimization and the analytical characterization studies of this dual enzyme biosensor will be discussed.

See more of this Session: Biosensor Devices
See more of this Group/Topical: Topical 9: Sensors

(203d) A Lipobead Microarray Made by Microfluidic Entrapment

Xiaoxiao Chen, Chemical Engineering, The City College of New York, The School of Engineering, New York, NY, Shahab Shojaei-Zadeh, Benjamin Levich Institute, New York, NY and Charles Maldarelli, Levich Institute and Chemical Engineering, The City College of New York, New York, NY

This presentation will describe the design of a bead-based, microfluidic array platform for the display of cell membrane sequestered receptors, to be used for the high-throughput, multiplexed screening of these receptors with extracellular target molecules. The design will be of multiple use, but our particular focus is to utilize the platform as a sensor for the detection and identification of bacterial exotoxins (pathogens) by displaying in the platform the cell membrane glycolipids which are selectively bound by the toxins as their first step in cell invagination.

Our array platform is a microfluidic cell fabricated by soft lithography out of polydimethyl siloxane (PDMS) with the surfaces of the fluidic pathways functionalized with polyethylene glycol (PEG) to prevent nonspecific adsorption of biomolecules. The cell consists of a single rectangular channel in which an obstacle course of capture elements (traps) have been arrayed to capture microbeads. The channel is approximately 60 microns in height and 1- 3 mm in width. The microbeads are 40 microns in diameter glass particles. Each trap consists of two concave elements to form a wine cup-shaped enclosure. Each element is separated from the other by a 25 microns gap in width, so that the flow can pass through the wine cup-shaped trap and a certain proportion of total streamlines can carry the lipobeads into the trap. Their capture is undertaken by streaming the microbeads through the cell, allowing the particles to be entrapped by the capture elements. In principle, the cell can be partially filled with one set of entrapped microbeads, spatially indexed by optical inspection, and then filled sequentially and indexed with other sets of beads to form a complete array of entrapped microbeads.

The microbeads are the hosts for the cell membrane receptors. In order to accommodate the strict requirement that the receptors need to be sequestered in a lipid bilayer to retain their selective and high affinity binding, lipid bilayers of DMPC are first formed around the particles to form lipobeads. This is undertaken by forming vesicles of DMPC by hydration of dried films of the phospholipid, and then fusing the vesicles to the silica surface. The receptors are sequestered in the bilayers by constituting the film with the receptors, or, in the case of delicate proteins, inserting them afterwards into the formed bilayer.

A functioning microfluidic array is constituted by trapping different sets of lipobeads, with each set displaying a different receptor. After entrapment, analyte solutions with labeled targets are flowed through the cell, and microbeads binding the labeled target can be identified using wide field fluorescence microscopy. The receptor on the microbead surface is identified by the array registrycompiled during the filling of the traps. The small size of the microbeads and the array allows for a miniaturization which is at least an order of magnitude smaller in size than can be accomplished with conventional microarrays set-up by using robotic fluidic dispensing to form spotted arrays.

In this presentation, we demonstrate the utility of this platform for multiplexed screening using biotin-streptavidin as an example of a selective biomolecular interaction. Biotin conjugated lipids – the receptor - are sequestered into a DMPC lipobead bilayer, and displayed in the array by fluidic entrapment. A second set of lipobeads, not containing the biotin "receptor" is also displayed by sequential entrapment. An analyte containing fluorescently labeled Neutravidin is then flowed through the cell. The Neutravidin selectively binds only to the microbeads containing biotin displayed on their surface as evidenced by the localized fluorescence on only the biotin lipobeads.

See more of this Session: Biosensor Devices
See more of this Group/Topical: Topical 9: Sensors

(203e) Loop Mediated Isothermal Amplification and Detection of Escherichia Coli Using Lateral Flow Strips

Ashish S. Yeri[1], Nilesh Dhamne[2], Abhay Vats[2] and Di Gao[1], (1)Chemical and Petroleum Engineering, University of Pittsburgh, Pittsburgh, PA, (2)Division of Pediatric Nephrology, Children's Hospital of Pittsburgh, Pittsburgh, PA

The need for rapid point-of-care diagnostics for infectious diseases has been steadily increasing globally, especially in developing countries. *E. Coli* is responsible for a large number of infections including water and food borne infections, urinary tract infection (UTI) and life threatening conditions such as neonatal meningitis and neonatal sepsis. Detection of *E. Coli* is generally done by bacterial cultures which usually takes 48-72 hours and requires skilled operators and expensive equipment and laboratories. Recently, loop mediated isothermal amplification (LAMP) technique has been developed which amplifies the DNA at around 63-65 °C with high specificity. The LAMP procedure utilizes 4 or 6 primers and can be performed on a regular water bath or a heating block maintained at the specified temperature. We developed a Nucleic Acid Lateral Flow (NALF) device to detect LAMP amplification products in an easy to read format. The detection of the LAMP *E. Coli* product is performed on a lateral flow strip which primarily consists of a nitrocellulose membrane with proteins/antibodies immobilized at specified locations. Labeled primers, namely biotin and FITC labeled outer primers are used in the amplification which takes about 30-60 minutes and the detection is performed using a sandwich type immunoassay. Results show a clear visible difference between the positive and negative samples and similar sensitivity and specificity when compared to PCR, making this assay a reliable, rapid and inexpensive diagnostic tool. An additional advantage could be the utility of this detection method in multiplexed assays where more than one type of *E. Coli* (such as toxigenic or hemorrhagic) or different bacteria (ie Klebsiella sp, Staphylococcus sp, etc) can be detected in as single assay and on a single strip in a point of care manner.

See more of this Session: Biosensor Devices
See more of this Group/Topical: Topical 9: Sensors

(203f) Highly Sensitive Detection of Antibody as An Indicator of Virus Exposure by Surface-Enhanced Raman Scattering (SERS) Spectroscopy

Jing Neng[1], Mark H. Harpster[1], William C. Wilson[2] and Patrick A. Johnson[1], (1)Chemical & Petroleum Engineering, University of Wyoming, Laramie, WY, (2)Center for Grain and Animal Health Research (CGAHR) USDA, ARS, Arthropod-Borne Animal Diseases Research Unit (ABADRU), Manhattan, KS

A rapid, highly sensitive, reproducible and point-of-care Surface-enhanced Raman scattering-based immunoassay system using gold nanoparticles (AuNPs) and dye-labeled protein A/G has been developed. In our system, the capture antigen is chemisorbed on gold nanoparticles (AuNPs), followed by incubating with the testing serum, which may contain the antibodies specific to the antigen. The Raman reporter is then added into the system. This reporter is fabricated by conjugating the dye molecule Malachite green to protein A/G which specifically recognizes and binds to the IgGs. The antibody links antigen covered AuNPs and Protein A/G conjugated reporters together to form a sandwich complex, bringing dye and Au surface in close proximity and initiating SERS effect. A characteristic signal of Malachite green indicates the existence of target antibody. This SERS-based immunoassay is applied to the detection of WNV specific antibody in immunosera, which is an indicator of exposure to WNV Virus. A quantitative curve is obtained by plotting the area of the SERS signal of Malachite Green at 1172 cm-1 and 1218 cm-1 versus the concentration of the immunosera against WNV antigen. In this immunoassay, the detection limit was 2 ng/ml of immunoserum which corresponds to at most 2 pg/ml of WNV specific IgG. This reported SERS-based immunoassay provides a rapid, highly sensitive, portable and cost effective viremia diagnostic tool as an alternative to time-consuming and labor-intensive methods restricted to analytical laboratories.

See more of this Session: Biosensor Devices
See more of this Group/Topical: Topical 9: Sensors

(203g) Non-Enzymatic Glucose Sensing Using Novel Metal Oxide-Metal Composite Nanofibers

Yu Ding, Ying Wang and Yu Lei, Department of Chemical, Materials and Biomolecular Engineering, University of Connecticut, Storrs, CT

Electrospinning technique has been applied to prepare metal oxide-metal composite nanofibers. Scanning electron microscopy and transmission electron microscopy were employed to characterize the morphologies of the nanofibers. X-ray diffraction and X-ray photoelectron spectroscopy were performed to investigate the crystal structure and compositions of the final product. The as-prepared metal oxide-metal composite nanofibers were then applied to modify the glassy carbon electrode to construct a non-enzymatic sensor for glucose detection in alkaline solution and a synergistic effect was achieved between the metal oxide and metal. Higher sensitivity, lower detection limit and wider linear range in glucose detection were achieved compared to pure metal oxide or metal modified electrodes.

See more of this Session: Biosensor Devices
See more of this Group/Topical: Topical 9: Sensors

(262a) Fluorescent Nanosensors for Glucose Detection In Vivo

Kevin J. Cash, Mary K. Balaconis, J. Matthew Dubach and Heather A. Clark, Biomedical Engineering Group, The Charles Stark Draper Laboratory, Cambridge, MA

Recent years have shown the incredible versatility and utility of polymeric nanosensors. Sensors have been developed for a wide variety of ionic analytes (e.g. sodium, potassium, pH) and have been applied for challenging research fields such as detection of sodium sparks in live cardiomyocytes. This field of research has yielded new knowledge on cellular metabolism and function. Additionally, these sensors can be applied in vivo, through implantation in or under the skin in animals; showing potential as a research tool for continuous monitoring of analyte concentrations without the need for direct samples to be taken from the animal.

In addition to ionic analytes, recent developments have enabled these polymeric nanosensors to be applied for the detection of nonionic analytes such as glucose through the use of boronic acid derivatives which bind to glucose. I will discuss the development of these sensors, including macroscale and nanoscale response; the optimization of components included in the sensors; and the design of the ratio of components to enable detection of glucose in the physiological range. Additionally, I will demonstrate the application of these nanosensors for the continuous tracking of blood glucose levels in vivo.

This work was supported by a grant from the National Institutes of Health National Institute of General Medical Sciences (R01 GM084366) and Internal Research and Development funding from The Charles Stark Draper Laboratory.

See more of this Session: Biosensor Devices II
See more of this Group/Topical: Topical 9: Sensors

(262b) Single-Molecule Detection of Nitroaromatic Compounds by Carbon Nanotube Optical Sensors

Daniel A. Heller[1], George W. Pratt[1], Jingqing Zhang[1], Nitish Nair[2], Adam J. Hansborough[1], Ardemis A. Boghossian[1], Nigel F. Reuel[1], Paul W. Barone[1] and Michael S. Strano[1], (1)Department of Chemical Engineering, Massachusetts Institute of Technology, Cambridge, MA, (2)Chemical and Biological Engineering, University of Colorado at Boulder

Polymer-encapsulated single-walled carbon nanotubes (SWNT) detect nitroaromatic compounds at the single-molecule with high selectivity and species recognition. A class of peptides from the bombolitin family not previously identified for nitroaromatic recognition allows the carbon nanotube around which they are encapsulated to report changes in their secondary structure, forming a 'chaperone sensor.' Analyte identification is afforded by spectroscopically differentiable changes of eight (n,m) SWNT species which generates a unique spectral fingerprint. An oligonucleotide $d(AT)_{15}$ enables high selectivity for trinitrotoluene to the exclusion of other nitroaromatics. These interactions are probed for the first time at the single molecule level, as stochastic adsorption and desorption events are optically reported by the SWNT as intensity or wavelength fluctuations. A novel dual-channel single-molecule imaging microscope is demonstrated to image quantized spectral shifts in real-time, resulting in the first single-molecule stochastic detection of solvatochromic events.

See more of this Session: Biosensor Devices II
See more of this Group/Topical: Topical 9: Sensors

(262c) Direct Electrochemistry and Electrocatalysis of Electrospun Single-Walled Carbon Nanotubes-Hemoglobin Composite Microbelts – towards the Development of Sensitive and Mediator-Free Biosensor

Yu Ding, Ying Wang and Yu Lei, Department of Chemical, Materials and Biomolecular Engineering, University of Connecticut, Storrs, CT

Single-walled carbon nanotubes (SWNTs) and hemoglobin (Hb) were co-electrospun to generate SWNTs-Hb microbelts, a novel composite material combining the advantages of excellent electron transfer property of SWNTs, electrocatalytic capability of redox Hb proteins, and highly porous structure of microbelts. FT-IR spectra confirmed that Hb in the microbelts kept its native conformational structure and Raman spectra demonstrated the successful incorporation of SWNTs in the microbelts. The direct electrochemistry of SWNTs-Hb composite microbelts was investigated and its application for electrocatalytic reductions and sensitive detections of three compounds, hydrogen peroxide, nitrite, and trichloroacetic acid, was also conducted. The comparison of the SWNTs-Hb composite microbelts modified electrode with the Hb microbelts modified electrode and other reported electrodes in the literature indicates that the incorporation of SWNTs into Hb microbelts can significantly enhance the direct electrochemistry of Hb and the SWNTs-Hb microbelts based biosensor has great potential application in the mediator-free detection of various analytes.

See more of this Session: Biosensor Devices II
See more of this Group/Topical: Topical 9: Sensors

(262d) Sensitive TNT Detection Using QCM Via Competition Assay

Jinghui Wang, Sibani Lisa Biswal and Laura Segatori, Chemical and Biomolecular Engineering, Rice University, Houston, TX

There have been many systems which have been designed for the detection of the explosive 2, 4, 6-trinitrotoluene (TNT). The current state of the art sensing schemes achieves high sensitivity, but low selectivity, which makes detection of TNT in a dirty environment difficult. Here we show how to combine a protein competition assay with quartz crystal microbalance (QCM) by using an anti-TNT antibody to detect TNT in small quantities and distinguish it from similar molecules. QCM is chosen due to its high sensitivity as a mass measurement sensor. ELISA (Enzyme-linked immunosorbent assay) is used to evaluate the competition assay for TNT detection and offer a comparison to the QCM results. Due to its label-free, real-time measurements, a TNT sensor based on a competition assay on QCM results in detection with high sensitivity and selectivity. Furthermore, this method could be also applied to detect other chemicals, such as environmental pollutants.

See more of this Session: Biosensor Devices II
See more of this Group/Topical: Topical 9: Sensors

(262f) SAW Induced Microfluidic Droplet Manipulation for Lab-On-a-Chip Systems

Reetu Singh[1], Subramanian Sankaranarayanan[2] and Venkat R. Bhethanabotla[1], (1)Department of Chemical & Biomedical Engineering, University of South Florida, Tampa, FL, (2)Center for Nanoscale Materials, Argonne National Lab, Argonne, IL

Abstract: Liquid droplet based microfluidic systems find extensive applicability in biological applications, such as carriers to target drug delivery to the intended diseased tissue, encapsulation agents for various biological entities, micro-reactors for rapid mixing and reaction of reagents, liquid reaction vessels for protein crystallization, biosensing, DNA analysis, sampling glucose concentrations in bodily fluids, to name a few. In droplet based microfluidics, discrete droplets are created via the use of two immiscible phases in microchannels. All applications require a great control on droplet size, shape, and distribution; mondispersed droplets are typically required which can be used as micro-reactors and for achieving hands-free contact of reagents in biological applications such as in Lab-On-a-Chip systems. Surface acoustic wave (SAW) devices, with hydrophobically modified surfaces, can be utilized for efficient microfluidic control and generation as well as actuation of droplets. In this work, we investigate droplet actuation using SAW devices as well as the acoustic streaming velocity fields generated in micro-liter size droplets as a result of interaction with SAW devices. A SAW device based, on YZ Lithium Niobate, in contact with a micro-liter sized droplet is modeled using a three dimensional bidirectionally coupled fluid-structure interaction model. The structure is simulated for a total of 100 nanoseconds (ns), with a time step of 1 ns. The excitation of the piezoelectric solid is provided by applying an AC voltage on the transmitter IDT fingers. The Rayleigh wave generated in the SAW device impinges on the fluid medium and imparts momentum generating fluid motion known as acoustic streaming. Our simulations indicate that when a droplet is placed on the SAW device, this acoustic streaming can actuate the droplet. Simulation derived fluid velocity profiles inside the droplet exhibit recirculation, thereby indicating an enhancement in mixing and agitation in droplets due to acoustic streaming, and thus underscoring their utility as micro-reactors. The influence of various parameters, namely fluid viscosity, applied voltage, and droplet size on the droplet actuation and acoustic streaming is studied using the developed fluid-structure finite element model. The role of acoustic wave focusing via the use of focused interdigital transducers, on droplet actuation as well as fluid velocity profiles within the droplet, is investigated as well. Results will be presented in detail.

See more of this Session: Biosensor Devices II
See more of this Group/Topical: Topical 9: Sensors

(262g) Paper Bioassays for Blood Typing

Mohidus Samad Khan, Department of Chemical Engineering, BioPRIA, Australian Pulp and Paper Institution, Monash University, Clayton, Australia, George Thouas, Division of Biological Engineering, Monash University, Clayton, Australia, Wei Shen, Chemical Engineering, Australian Pulp and Paper Institution, Monash University, Clayton, Australia, Gordon Whyte, School of Rural Health, Faculty of Medicine, Nursing & Health Science, Monash University, Australia and Gil Garnier, Department of Chemical Engineering, BioPRIA, Australian Pulp & Paper Institute, Monash University, Melbourne, Australia

Agglutinated blood transports differently onto paper than stable blood with well dispersed red cells. This difference was investigated to develop instantaneous blood typing tests using specific antibody-antigen interactions to trigger blood agglutination. Two series of experiments were performed. The first related the level of agglutination and the fluidic properties of blood on its transport in paper. Blood samples were mixed at different ratios with specific and non-specific antibodies; a droplet of each mixture was deposited onto a filter paper strip and the kinetics of wicking and red cell separation were measured. Agglutinated blood phase separated, with the red blood cells (RBC) forming a distinct spot upon contact with paper while the plasma wicked; in contrast, stable blood suspensions wicked uniformly. The second study analyzed the wicking and the chromatographic separation of droplets of blood deposited onto paper strips pretreated with specific and non-specific antibodies. Drastic differences in transport occurred. Blood agglutinated by interaction with one of its specific antibodies phase separated, causing a chromatographic separation. The red cells wicked very little while the plasma wicked at a faster rate than the original blood sample. Blood agglutination and wicking in paper followed the concepts of colloids chemistry. The immunoglobin M antibodies agglutinated the red blood cells by polymer bridging, upon selective adsorption on the specific antigen at their surface. The transport kinetics was viscosity controlled, with the viscosity of red cells drastically increasing upon blood agglutination. 3 arms prototypes were investigated for single-step blood typing.

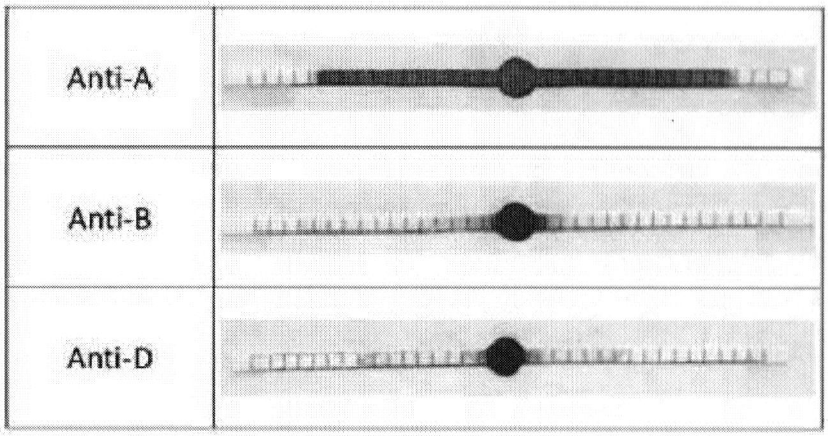

(a) Blood Typing Technique-1: B+ blood group

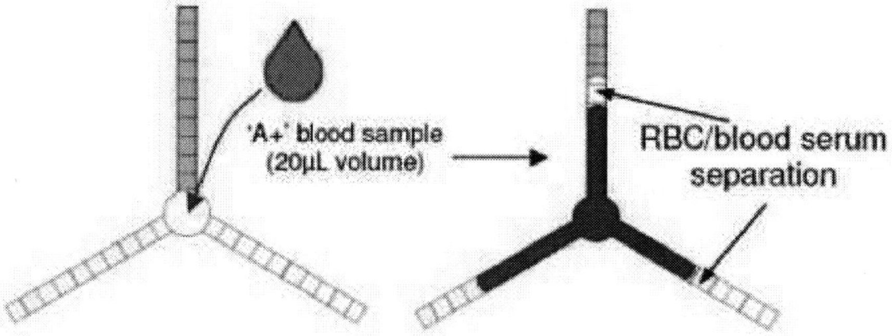

(b) Blood Typing Technique-2: A+ blood group

See more of this Session: Biosensor Devices II
See more of this Group/Topical: Topical 9: Sensors

(316a) Hydrogel-Based Protein Array for Quantification of EGFR Activity in Cancer Cellular Extracts

Gargi Ghosh, Department of Chemical and Biological Engineering, University of Wisconsin-Madison, Madison, WI, Andrew G, Lee, University of Wisconsin-Madison, Madison, WI and Sean P, Palecek, University of Wisconsin-Madison, Madison, WI

Epidermal growth factor receptor (EGFR) signaling pathways have been implicated in playing an important role in cancer development and progression. This led to the development of therapeutics directed against the tyrosine kinase domain of EGFR. However, therapeutics targeted against EGFR either have failed to reproduce promising preclinical model results in clinical settings or have only been successful in a cohort of cancer patients bearing somatic mutation in EGFR kinase domain. The failure can be partly attributed to incomplete assessment of EGFR status in cancer, inappropriate drug dosage and /or development of drug resistance during treatment. One way to address this problem is to develop a customized targeted treatment. A patient-customized, predictive diagnostic that quantifies the effects of specific anti-EGFR therapies may improve outcomes in cancers where EGFR plays a mechanistic role. Nonetheless, major challenges in tailoring patient-specific treatment regimens involve predicting the most effective inhibitors and monitoring for treatment progression. We believe any system capable of directly measuring the EGFR activity will provide a more accurate report of the disease state than the conventional methods which involve quantification of EGFR at DNA, RNA, or protein levels. However, technical challenges in developing such a tool involved specific quantification of EGFR activity in the background of multiple tyrosine kinases present in cancer cellular extracts and obtaining accurate measurements from small biopsies. Here we report the development of a novel hydrogel based protein array for quantifying the activity of EGFR in cancer cellular extracts. Acryl-labeled EGFR substrate (glutathione S-transferase – Eps15 (GST-Eps15) fusion protein) is polymerized within a polyacrylamide hydrogel that is immobilized to a surface. Incorporation of the substrate within the hydrogel network helps maintain the native form of the protein by providing a hydrated environment and increases the accessibility of the substrate to kinases in the cell lysate. These characteristics provide a probe with high binding capacity and sensitivity, ultimately improving signal-to-noise ratio. The extent of substrate phosphorylation upon incubation with lysates of cancer cells corresponded to the difference in activation status of EGFR in different cancer cell lines. Significant EGFR upregulation was detected in a mixture containing 7% EGFR-overexpressing (breast cancer cell line, MDA-MB-468) cell lysate diluted in lysate from a (breast cancer cell line, MDA-MB-453) cell line expressing low levels of EGFR. Approximately, 2000 cells are required for quantification of EGFR activity. Additionally, the ability of the protein array to quantify the extent of inhibition of EGFR activity by different tyrosine kinase inhibitors establishes the potential of this protein-acrylamide copolymer hydrogel array to screen for the most promising therapeutics for individual patients and monitor treatment progression.

(316b) Protein Microarrays From Zwitterionic Carboxybetaine for Highly Sensitive Early Cancer Diagnostics Using SPR Imaging

Norman D. Brault, Qiuming Yu and Shaoyi Jiang, Chemical Engineering, University of Washington, Seattle, WA

Research has shown the extraordinary heterogeneity of cancer (numerous mutated and deregulated genes, abnormal expression of microRNAs, and genetic instability). This has lead to the use of high-throughput antibody-based biosensor platforms which can detect multiple disease biomarkers simultaneously in order to enhance the sensitivity and accuracy of diagnostic assays. The ability to sensitively detect these target analytes (at or below nanogram/milliliter quantities) from the complex environment of human blood serum or plasma has severely hindered the biomarker development and verification process using biosensors due to the presence of non-specific protein adsorption. In this work, antibody arrays were created using a simple yet robust zwitterionic poly(carboxybetaine methacrylate) (pCBMA) surface platform. The functionalization capabilities combined with the ultra-low protein fouling properties both before and after effective immobilization of biorecognition elements made pCBMA highly desirable. Using standard pin-spotting techniques on the pCBMA films, arrays containing antibodies to three cancer biomarkers were analyzed using a highly sensitive self-referencing surface plasmon resonance (SPR) imaging sensor with polarization contrast. The immobilization efficiency and reproducibility as well as subsequent biofouling, antibody activity, and stability in complex media were evaluated. Finally, the simultaneous detection of the three biomarkers directly from undiluted human serum was demonstrated.

See more of this Session: Biosensor Devices III
See more of this Group/Topical: Topical 9: Sensors

(316c) Using Functionalized DNA Nanostrands as Sensitive Biosensors

Bo Yu[1], Nan-Rong Chiou[2], Pouyan E. Boukany[3], Paul Bertani[3], Wu Lu[3] and L. James Lee[4], (1) Department of Chemical and Biomolecular Engineering, The Ohio State University, Columbus, OH, (2) Chemistry, The Ohio State University, Columbus, OH, (3)Nanoscale Science and Engineering Center for Affordable Nanoengineering of Polymeric Biomedical Devices, The Ohio State University, Columbus,, OH, (4)The Ohio State University, Columbus, OH

Current biology research and detection are based on averaged information from a relatively large sample size, which limits our investigation of rare cell behavior (e.g. cancer stem cells) and the possibility of early disease diagnosis. Furthermore, most of existing methods require more than 106 cells and long assay times to detect proteins or nucleic acids of interest because of the resolution limit of the detection methods. It would be highly valuable to have new techniques capable of biomolecular analysis at the individual cell or even single molecule basis. The large aspect ratio of the nanoporous materials offers a potential for sensing biological species at a single molecular level with fast response. The sensitivity of such depends strongly on the size of the nanochannels, i.e., smaller the pore diameters leading to more sensitive detection. However, it is very difficult and expensive to fabricate nanochannel(s) with the pore sizes smaller than 5 nm without the substantial defects from the top-down approach. Recently, we developed a novel DNA Combing and Imprinting (DCI) process to fabricate adjustable nanochannel arrays using functionalized dsDNA as the template, called DNA nanostrands. These DNA nanostrands were simply generated on the micro patterned PDMS stamp in a molecular combing process by fast de-wetting. They were then transferred onto a solid substrate such as a glass plate or a silicon wafer and cast with a liquid polymeric resin to from a confined nanoscopic elements. By conjugating various complimentary nucleic acids, antibody molecules or nanoparticles onto the dsDNA chains before DCI, the gallery spacing in the DNA nanostrands can be adjusted and the surface of DNA nanostrands can be functionalized. Such configuration allows us to study ion and molecular transport in the nanoscaled domain as well as label-free detection of the nucleic acid and antigen by electrophoresis. In this study, we conjugated complementary oligo nucleic acids on calf thymus DNA, and investigated the selectivity and sensitivity of this biosensor for oligodeoxyribonucleotide (ODN) and microRNA detection.

See more of this Session: Biosensor Devices III
See more of this Group/Topical: Topical 9: Sensors

(316d) Surface-Enhanced Raman Spectroscopy (SERS) for Ultratrace Viral DNA Detection by Magnetically Concentrating Raman Active Gold Nanoparticles

Hao Zhang, Department of Chemical & Petroleum Engineering, University of Wyoming, Laramie, WY and Patrick A. Johnson, Chemical & Petroleum Engineering, University of Wyoming, Laramie, WY

A practical format for SERS bioassays is to use activated nanoparticles in solution. By performing the assay in solution, one can use high laser power to interrogate the sample without concern of burning the sample (as with solid-phase assays). Moreover, the biomolecules conjugated nanopaticles can be utilized to target pathogen DNA, Virus antigen with high efficiency. In our previous study, we report the demonstration of a SERS-based solution assay for the indirect capture of a viral DNA sequence using a target sequence derived from the West Nile Virus (WNV) genome. However, generally, solution-based assays are not very sensitive since the SERS particles are in low concentrations, not easily interrogated by a narrow laser beam, and freely adrift in solution.

In this study, we propose to overcome the shortcomings in Raman solution assay by developing methods to consistently fabricate highly active SERS reporter particles and to amplify the SERS signal (ie. increase analyte detection sensitivity) through an innovative paramagnetic capture method. A 53-mer sequence derived from the West Nile Virus (WNV) genome was chosen as the target DNA. Two DNA probes complementary to one-half of the target DNA were generated and conjugated to magnetic nanoparticles (MNPs) and Raman active gold nanoparticles (GNPs). After hybridization with target DNA, two nanoparticles form combination structures which can be magnetically collected and located. This concentrated combination of Raman Active gold nanoparticles and paramagnetic particles was will amplify the SERS signal in two ways: 1) by magnetic concentration of the bound analyte and Raman reporter tag and 2) by increasing the content of nanostructured Au and therefore increasing the probability of formation of SERS "hot spots". This concentration method will increase the types of sample matrices which can be examined. For example, very "dirty" or highly dispersed samples can be assayed since the matrix material is eliminated from a magnetically concentrated spot positioned within the focal point of the interrogating laser.

See more of this Session: Biosensor Devices III
See more of this Group/Topical: Topical 9: Sensors

(316e) Detection of Double and Single-Stranded DNA by 5CB

Stephanie M. Malone, Chemical and Biological Engineering, University of Colorado at Boulder, Boulder, CO and Daniel K. Schwartz, Chemical and Biological Engineering, University of Colorado, Boulder, CO

We have investigated the disruption of azimuthal alignment of the liquid crystal 5CB by double and single-stranded DNA at the solid-liquid crystal interface. A liquid crystal-based microarray has the potential to reduce expenses when compared to similar fluorescence-based systems, both with equipment capital and the elimination of the chemistry behind fluorescent labeling. Simplification of imaging equipment to a simple crossed polarized system also could increase availability of rapid genetic screening, expanding availability to diagnostic applications. Ultimately, hybridization detection could lead to a liquid-crystal based microarray. This approach uniquely differentiates between double or single-stranded DNA by detecting the inherent structure of DNA rather than detection of a molecular labeling agent. Sensitivity to the native state of DNA eliminates detection problems through nonspecific DNA binding to a sensor. We observed a minimum threshold for sensitivity has been observed to be 200 molecules per mm^2 with theoretical sensitivity as low as 30 molecules per mm^2. Combined polarized and fluorescence microscopy is used to simultaneously observe physical alignment of DNA as well as liquid crystal alignment. Anchoring energies of double and single-stranded DNA are also compared to further quantify disruption of the director field.

See more of this Session: Biosensor Devices III
See more of this Group/Topical: Topical 9: Sensors

(316f) Rapid and Sensitive Detection of Salmonella Using An Integrated Microfluidic Chip and Real-Time PCR

Tao Geng, Department of Agricultural and Biological Engineering, Purdue University, West Lafayette, IN, Krishna Mishra, Department of Food Science, Purdue University, West Lafayette, IN, Arun K. Bhunia, Food Science, Purdue University, West Lafayette, IN and Chang Lu, Chemical Engineering, Virginia Tech, Blacksburg, VA

Salmonella is one of the leading causes of food-borne illness. Conventionally, the presence of Salmonella in food products is routinely tested by a lengthy and laborious procedure including non-selective pre-enrichment, selective enrichment, selective plating, and subsequent biochemical and serological confirmation, which requires 4 and 6 days for negative and positive samples, respectively. To overcome these issues, we developed a strategy that integrates sample pretreatment steps including cell lysis and DNA purification on a single microfluidic chip in order to reduce the processing time and the number of steps for rapid detection of pathogens. In addition, immunomagnetic separation (IMS) technique and real-time PCR assay were also incorporated in our methods to achieve both high sensitivity and specificity of detection. The target bacterial cells were initially separated from crude food samples using Dynabeads® anti-Salmonella. The bead-Salmonella complex was then delivered into the microfluidic chip and captured by a NdFeB permanent magnet. Subsequently, the bacterial cells could be lysed by electrical pulses with high intensity (1000-1500 V/cm) within only 1-3 minutes. The released DNA was selectively adsorbed onto magnetic ChargeSwitch® beads held by a magnet in the down-stream of microchannel, which greatly removed PCR inhibitors from samples. We were able to detect Salmonella with a high sensitivity level, and the total assay time is only about 6 hr. The results suggest that our approach provides a new tool for rapid, sensitive, accurate and cost-effective identification of pathogens.

See more of this Session: Biosensor Devices III
See more of this Group/Topical: Topical 9: Sensors

(316g) Tethered Bilayer Lipid Membranes with Pre-Designed Nanopores for DNA-Membrane Interaction Analysis

Kwang Joo Kwak, Chemical and Biomolecular Engineering, The Ohio State University, Columbus, OH, Xuejin Wen, Electrical and Computer Engineering, The Ohio State University, Columbus, OH, Xin Hu, NSEC center for Affordable Nanoengineering of Polymer Biomedical Devices (CANPBD), The Ohio State University, Columbus, OH, Xinmei Wang, NSEC Center for Affordable Nanoengineering of Polymeric Biomedical Device (CANPBD), The Ohio State University, Columbus, OH, Gintaras Valincius, Chemistry and Bioengineering, Institute of Biochemistry, Vilnius 08662, Lithuania, David J. Vanderah, Center for Advanced Research in Biotechnology, National Institute of Standards and Technology, Rockville, MD, Wu Lu, Computer and Electrical Engineering, Ohio State University, Columbus and L. James Lee, Ohio State University, Columbus, OH

By selecting cell membrane lipids and adding proper protein and sugar molecules, the tethered bilayer lipid membrane (tBLM) with well-defined nanopore array can serve as a useful model for the investigation of various biological lipid structures and their interactions with other biomolecules. Examples include cell membrane-DNA interactions, cell membrane-lipoplex interactions, cell membrane-exosomes interactions and gene delivery by electroporation. The tBLM with well-defined nanopores was demonstrated on the SiO2 surface with pre-drilled nanoholes. The nanohole array with ~10,000 holes was prepared by e-beam lithography for electrochemical impedance spectroscopy (EIS) measurements. A thin self-assembled monolayer (SAM) was then formed on the Au layer. The tethered BLM was prepared by incubation with a diphytanoylphosphatidylcholine (DPhyPC) solution of 10 mM concentration in ethanol for ~10 min and subsequently injection of PBS buffer within ~10 s. The atomic force microscopy (AFM) image with the nanopores on a tBLM was observed in the liquid environment. The nanoholes with a diameter of 150 nm were drilled through the Au-coated SiO2 by e-beam lithography, while the observed nanopore size was around 100 nm. Since the lipid bilayers could also form on the sidewalls of the nanowells, the nanopore size was smaller than the nanohole size as measured by AFM. The tBLMs with different pore size and pore density were also prepared. The Cole-Cole EIS spectra of SAMs and tBLMs with and without nanoholes were measured and compared to the finite element method (FEM) simulation based on a 3-layer circuit model. For the SAM with nanoholes, the top gold surface and the sidewalls of nanoholes were modified with the mixed SAMs. Although the SAM layer was essentially defect free, the EI spectra showed some difference. For the tBLMs, the presence of nanoholes made the high frequency semicircle of the EI spectrum incomplete, similar to that observed when the densely-packed tBLM turned into a loosely-packed tBLM (lp)tBLM because of the reduced DPhyPC concentration. Unlike the (lp)tBLMs, this tBLM with well-defined nanopore array showed very stable and repeatable EI spectra. Several plasmid and linear DNAs used in gene therapy such as vector pGFP and pCAG2LMKOSimO are placed on top of the porous tBLM. The AFM is used to measure the morphology of DNA-pore interaction, while EIS is used to quantify the DNA-pore dynamics. We have designed a nanopore electrophoresis set-up to carry out these experiments.

See more of this Session: Biosensor Devices III
See more of this Group/Topical: Topical 9: Sensors

(390a) A Microfabricated Carbon Dioxide Sensor for Portable Applications

Brian A. Rosen[1], Amin Salehi-Khojin[1] and Rich Masel[2], (1)Department of Chemical and Biomolecular Engineering, University of Illinois at Urbana-Champaign, Urbana, IL, (2)Department of Chemical and Biomolecular Engineering, University of Illinois, Urbana-Champaign, Urbana, IL

The reliable detection of carbon dioxide emissions is a key step towards the ability to mitigate CO2 levels in the atmosphere. At present, many CO2 sensors exist in the literature. Mixed oxide capacitors work very well, but require watts of power to maintain the sensor at the operating temperatures, typically 400-500 C. Infrared CO2 sensors also work well, but the lasers usually require a few hundred milliwatts of power to give reliable results. Such sensors are quite useful when attached to the power grid, but they do not provide the possibility of a long term portable unattended CO2 sensor. In this paper we present a new design for a microfabricated CO2 sensor that operates at room temperature, with very low power input. The device uses the selective adsorption of CO2 in a room temperature ionic liquid followed by an electrocatalytic reaction to detect CO2. We find that the device is able to measure normal atmospheric levels of CO2 with high reliability using less that 10 microwatts of power. This advance will enable a new class of portable CO2 sensors to be constructed.

See more of this Session: Catalytic, Environmental, and Industrial Gas Sensors
See more of this Group/Topical: Topical 9: Sensors

(390b) Ammonia Gas Sensing by Zeolite-Fiber Optic Device: Effects of Surface Modification and Operation Temperature

Xiling Tang, Zhi Xu, Seok-Jhin Kim and Junhang Dong, Department of Chemical and Materials Engineering, University of Cincinnati, Cincinnati, OH

A ZSM-5 zeolite thin film is synthesized on a long period fiber grating (LPFG) optical device. This zeolite-fiber device is used for ammonia sensing by monitoring its resonance wavelength which shifts due to the change of the zeolite refractive index caused by adsorption of molecules in the zeolite. The selectivity, sensitivity and speed of response of the zeolite coated-LPFG to an analyte gas are determined by the equilibrium and transport properties of gas adsorption in the zeolite film. In this study, the ZSM-5 zeolite film is modified to increase its surface acidity that dramatically improves the sensitivity and selectivity for ammonia gas detection. The effect of operation temperature on the sensitivity, selectivity/specificity, and response time is experimentally investigated.

See more of this Session: Catalytic, Environmental, and Industrial Gas Sensors
See more of this Group/Topical: Topical 9: Sensors

Infrared Optical Imaging for Advanced Gas Leak Detection

Anisa Safitri[1], Xioadan Gao[2] and M.S. Mannan[2], (1)Mary Kay O'Connor Process Safety Center, Artie McFerrin Department of Chemical Engineering, College Station, TX, (2)Chemical Engineering, Artie McFerrin Department of Chemical Engineering, College Station, TX

Abstract

Infrared imaging technique is used in industry as a method to detect fugitive leaks from equipment and pipeline systems. Optical imaging is considered as a smart LDAR (Leak Detection and Repair) because it can scan a large number of equipment in relatively short time compared to detection using Total Vapor Analyzer (TVA) or 'gas sniffer'. In addition, the ability of infrared optical imaging system to visualize the gas plume which is not visible to naked eyes offers another advantage. However, this novel technique poses a lot of challenges in its application due to many uncertainties related to the sensitivity of the camera and factors which may affect measurement. Infrared imaging technique has been used in this research to detect methane gas leak from pipelines and monitor LNG plume from LNG spill. In this work, some significant factors affecting measurement such as gas emissivity, atmospheric attenuation, and stimulated radiation from other objects than the target are being evaluated. Furthermore, potential use of infrared imaging technique for methane gas emissions estimation is proposed in this research. This is carried out by assessing the sensitivity of the infrared camera during monitoring the gas release in order to obtain the minimum detectable gas concentration that still can be observed by the camera under real meteorological conditions. From this test, the correlation of mass flow rate and distance to minimum detectable concentration will be withdrawn. Prior to the test, discharge and dispersion simulation of methane gas at various pressures, temperatures and leak sizes is performed to calculate the gas release rate and predict the downwind concentration of methane gas. Several types of infrared imaging technique also have the capability as a non-contact temperature measurement and thus they can provide a spatial temperature distribution of a target object. This facility is used in this research to demonstrate the temperature profile of LNG gas plume in case of accidental spill of LNG on the ground. However, there is a high discrepancy of the cloud temperature measured using infrared camera to the thermocouple reading. This research has been able to identify the most significant uncertainty which comes from unspecified emissivity of the LNG cloud. The methane gas emissivity is not included in the detector's algorithm and therefore the apparent target temperature shows much higher value than the actual one. In this study, the methane gas emissivity as a function of temperature at different optical depth is analyzed using band absorption model.

Introduction

Potential use of infrared optical imaging for hydrocarbon detection and visualization has been recognized and found useful in industry. The advantage of utilizing infrared imaging system is that it can detect process anomaly such as gas release from a leak or spill of volatile material without having to be in the proximity to the leaking flammable material. In addition, the ability of the infrared camera to visualize gas or vapor can be used to predict the amount of the gas being emitted.

Infrared camera detects the absorptive and emissive characteristics of several gases. The use of infrared optical imaging for fugitive methane gas detection and visualization had been demonstrated earlier in several literatures. In 1991, Gross et al detected methane gas leaks from buried, leaky pipelines was monitored using a staring, high resolution, PtSi Schottky barrier focal plane array infrared camera with 256 x 256 pixels and cooled to 77 K. The test was conducted under a controlled environment. Therefore influence of wind-force and moisture level was considered negligible. A 30-cm diameter was observed; methane concentration as low as 0.03 % can be detected from the thermogram.

Very similar camera system had been applied to monitor methane leaks from pipelines under realistic condition by Schulz et al where the infrared optical imaging was used to monitor leaks from a natural gas pipeline buried 1.2 m underground. The camera was able to identify a leak as low as 2 L/min, under moderate weather conditions, and with a 2 m/s wind speed. The cloud size observed in this study was 50 cm in diameter. *(Schulz et al., 2008)*

In the oil and gas industry, optical infrared imaging technique has been utilized by Total E&P Indonesie survey any VOC releases from equipment and pipelines. Survey of potential hydrocarbon release is important to protect people and to eliminate the environmental impact of hydrocarbon release.

Although infrared imaging can detect leaks accurately, it is not capable to provide quantitative measurements of the emissions rate or the gas concentration. Quantitative measurements of emission rates and concentration are preferable in order to estimate the total emissions released from the leak which is important in determining the hazardous distance or region in case of release of flammable material.

Methodology

The methodology for estimating methane gas emissions rates from natural gas leak or spill of LNG given in Figure 1. In the event a leak occurs in a natural gas pipeline, a certain amount of gas will be released and dispersed. The emission rate depends on the differential pressure between the upstream pressure and ambient pressure, the size of the leak, and the temperature of the gas inside the pipeline. The concentration of the gas as it is dispersed can be estimated using Gaussian dispersion model and the correlations between leakage rates and plume size, i.e. distance and plume height to minimum detectable downwind concentration for methane gas, can be predicted.

The effect of humidity, atmospheric content and ambient temperature to the atmospheric transmittance were also studied in order to obtain the true temperature profile of the dispersed gas. Emissivity factor is important to determine the amount of radiation received by the camera detector. The amount of water vapor in terms of the partial pressure of water vapor in the mixture of atmospheric gases is determined from the relative humidity and saturation vapor density where the infrared camera is used to detect hydrocarbon gases.

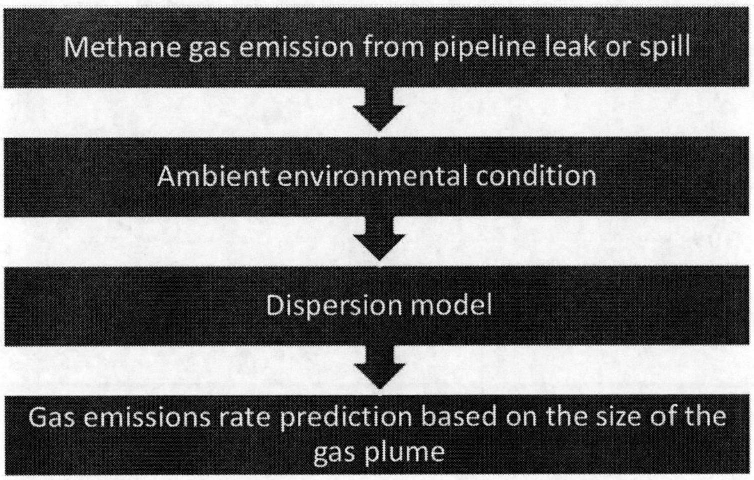

Figure 1. Methodology to estimate methane gas emissions rate

The composition and partial pressure of the atmospheric gas mixture can be determined at various humidity and ambient temperature. The emissivity of the mixture of atmospheric gas and flammable gas can be calculated using the weighted sum of grey gas model at various compositions of atmospheric gases.

Results and Discussion

Methane gas release from a leak in a pipeline

The discharge rate of gas coming out of a leaky pipeline depends on the gas temperature as well as the differential pressure between the gas inside the pipe and the atmospheric pressure. At upstream pressures higher than 187 kPa, the gas becomes critical and the gas velocity reaches choked velocity. As the flow of gas becomes choked, the gas mass flux keeps due to the increase in the gas density. A moderate temperature decrease of the natural gas inside the pipeline will not affect the amount of gas discharged from the leak source.

In the natural gas transportation and distribution system, gas is usually transported at ambient conditions, except for underground pipelines in which the temperature of the gas is a little lower than the ambient temperature during warm weather and higher than ambient temperature during cold weather. The mass discharge rate of the gas from 0.1 inch leak at various pressures and temperatures is given in figure 1. From the figure, it can be implied that the mass flow rate highly depends on the upstream pressure, but the effects of a slightly increasing or decreasing temperature is insignificant to the gas mass flow rate.

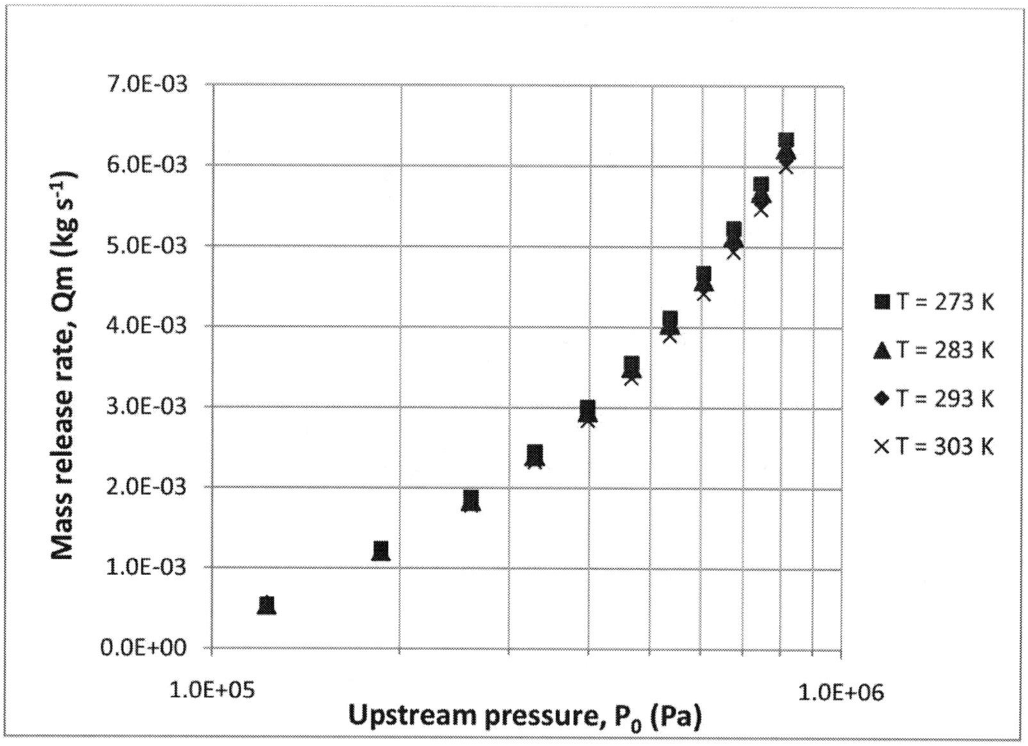

Figure 1. Mass discharge rate of methane gas at different pressures and temperatures from 0.1 inches leak source.

As the gas flow turns into choked the exit temperature becomes constant although the pressure in the upstream is increased. Figure 2 shows the gas temperature profile at the point of release for different initial temperatures. The exit parameters, i.e. pressures and temperatures, are crucial to determine the phase of methane gas being released in order to identify if any condensation had occurred due to the expansion of gas from a pressurized source to atmospheric conditions.

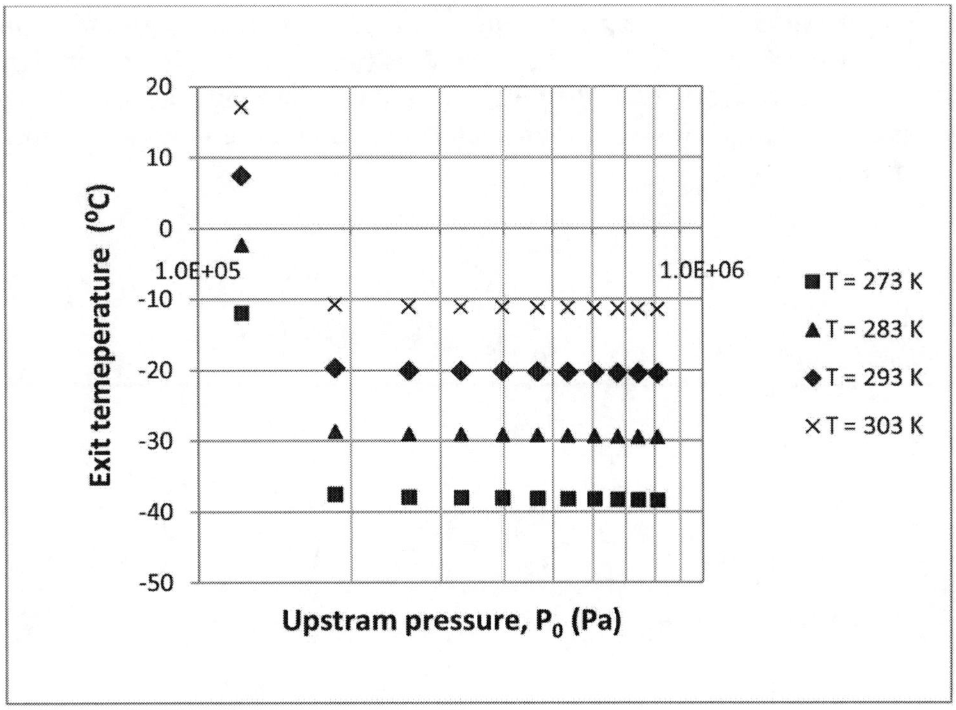

Figure2. Exit temperatures of methane gas at various upstream pressures and gas initial temperatures.

The density of gas as a function of the pressure and temperature can be seen in Figure 3. Upon increasing the temperature, the density decreases because the molecules are more spread out at higher temperature.

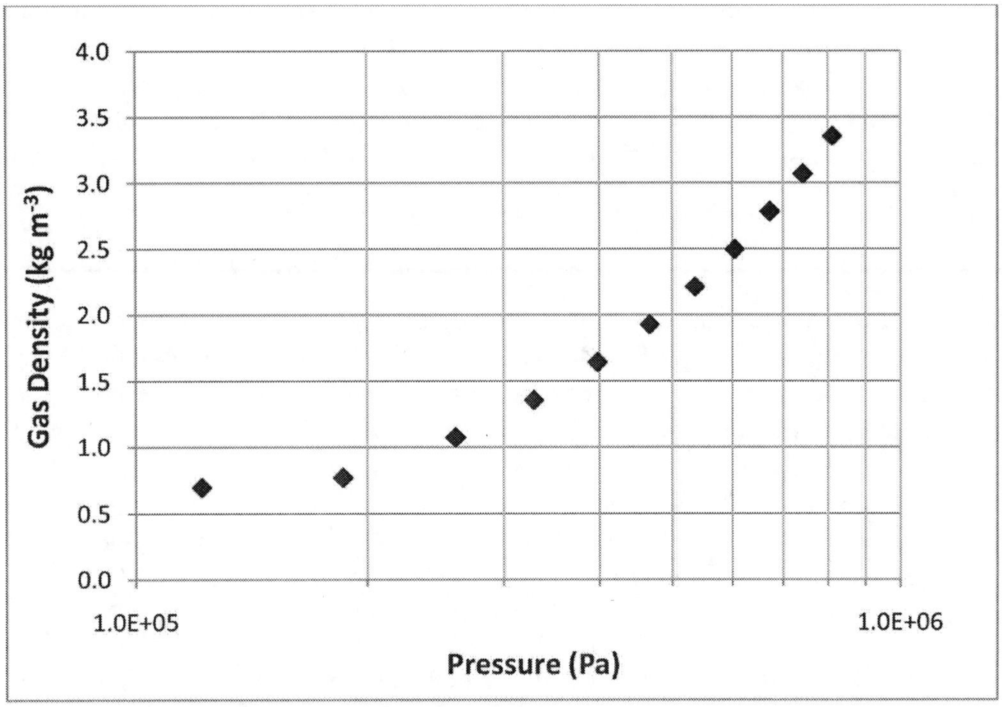

Figure3. Pressure-dependant gas density of methane

The volumetric flow rate of the gas exiting from 0.1 inch hole at various pressures and temperatures is shown in Figure 4. The volumetric flow rate of the gas increases significantly at non-choked condition and stabilizes as the velocity of the gas exiting reaches the sound speed. Increasing pressure and temperature does not change the volumetric flow rate dramatically for a hole with the same diameter.

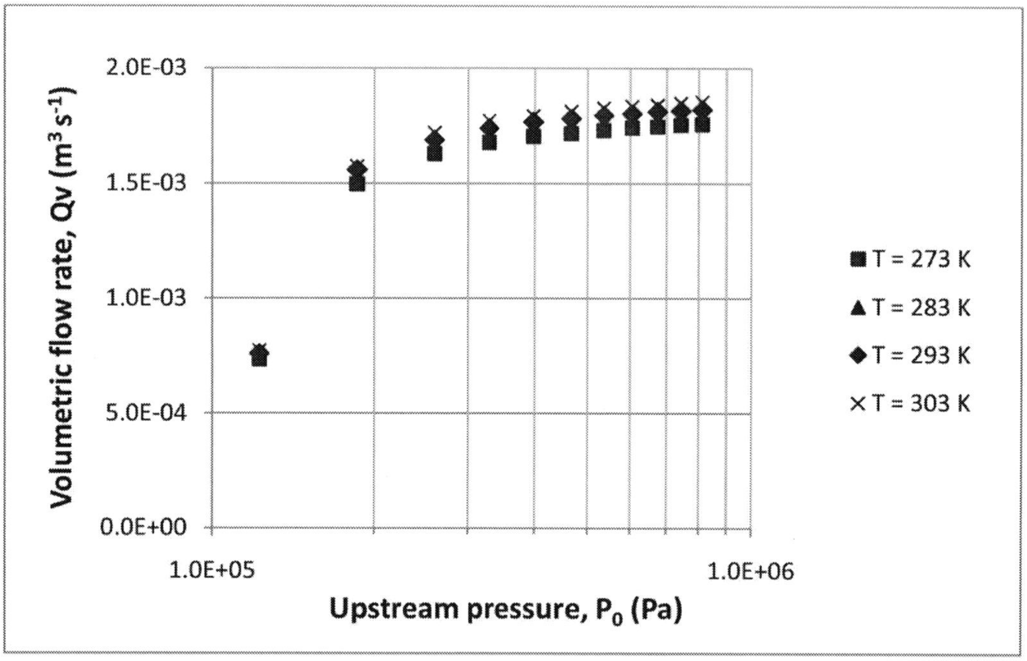

Figure4. Volumetric flow rate of methane gas released from 0.1 inch diameter

The mass flow rate can be calculated for different size of leak. The mass flow rate is the product of gas volumetric flow rate and gas density. The mass release rate increases as leak areas become larger. In the real gas flow, releases from a large leak will result in a decrease in upstream pressure; however, in this calculation a large inventory of gas inside the chamber is assumed. In this case, the stagnant upstream condition is assumed and a decrease in upstream pressure is not taken into account.

Methane gas dispersion

After the source or discharge term is specified, the dispersion of gas is simulated using the generalized Gaussian model. Pasquill-Gifford stability class is applied to calculate the concentration profile along the downwind and vertical direction. In the dispersion calculation, the mass discharge rate is assumed to be constant and the gas in the pipelines is at stagnant condition. Using this assumption, the concentration downwind and crosswind is solely a function of meteorological conditions. In this work, the dispersion of methane gas is calculated for three different weather conditions: (i) stable weather condition (Stability Class F) with 2 ms^{-1} wind speed, (ii) neutral weather condition (Stability Class D) with low wind speed (2 ms^{-1}), and (iii) neutral weather condition (Stability Class D) with medium wind speed (5 ms^{-1}).

The sensitivity of infrared imaging to detect methane gas concentration at real weather conditions is up to 600 ppm based on previously published works. Therefore, the gas emission rate is calculated based on the size of the cloud up to this concentration called the minimum detectable concentration. Figure 5 demonstrate the downwind distance to the minimum detectable concentration as a function of gas mass release rate at various weather conditions. From that figure it can be inferred that the downwind distance is highly affected by the wind speed. However, the change due to a different weather stability is not very significant for Class F and Class D. Based on the prediction using the Gaussian model, the downwind distance to 600 ppm of methane gas is increased by approximately a factor of two when the wind speed goes up from 2 ms^{-1} to 5 ms^{-1}.

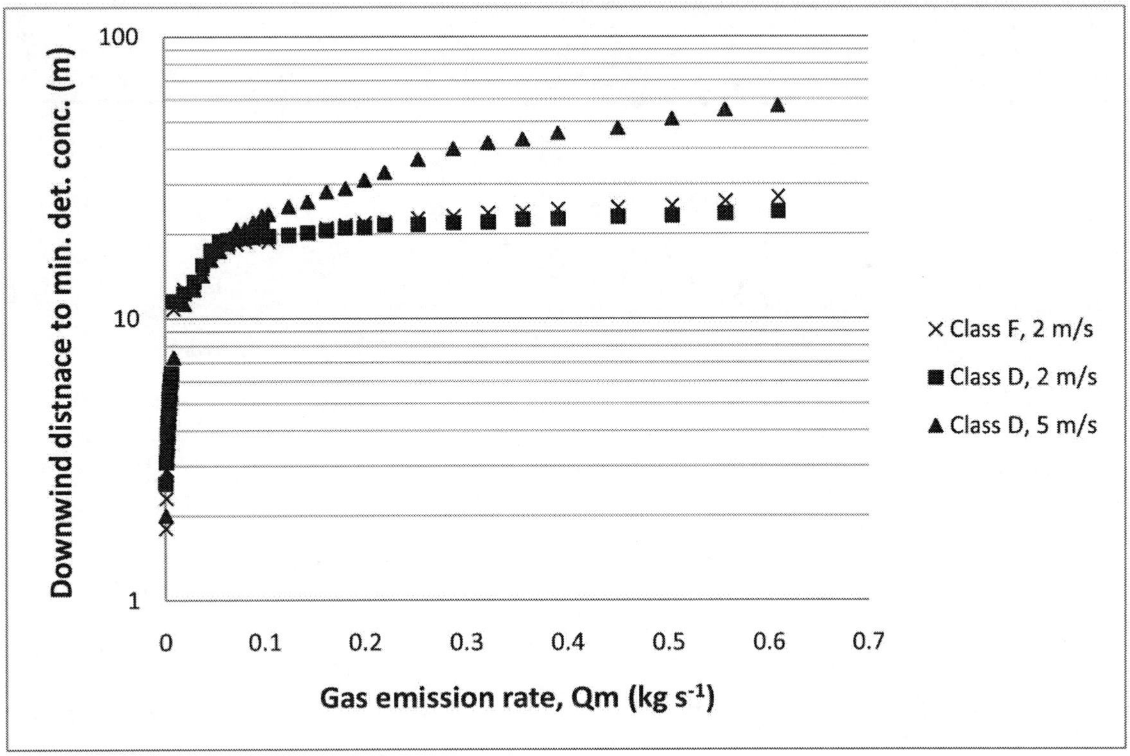

Figure 5. Downwind distance to minimum detectable concentration for different stability classes.

Application of infrared camera to determine temperature profile of methane gas

The aim of this research work is also to predict the emissive power of a mixture of hydrocarbon gas and air in the atmosphere by determining the emissivity of the gas mixture quantitatively. Humidity, ambient temperature, optical path length are among the factors which are affecting the performance of infrared camera in visualizing a target hydrocarbon gas. Water vapor and carbon dioxide are two gases in the atmosphere which have active absorbance within the mid-wave IR range in addition to the presence of methane gas. The significance of these gases in absorbing and emitting the infrared energy depends on their partial pressures in the atmosphere. The amount of water vapor in the atmosphere is usually presented in terms of relative humidity.

Partial pressure of water vapor in the atmosphere is increased with humidity level. Figure 6 shows the composition of the gases in the atmosphere at various relative humidity and ambient temperature. From the figure, it can be inferred that increasing ambient temperature by 10 K from 273 K to 283 K will result in the increase of the partial pressure of water vapor by a factor of approximately 3. Increasing ambient temperature by 20 K from 273 K to 293 K will increase in concentration by a factor of 5, and if the ambient temperature is increased by 30 K from 273 K to 303 K, partial pressure of water vapor will increase by a factor of 12. This indicates that ambient temperature has a considerable effect to determine the partial pressure of water vapor in the atmosphere. As the relative humidity rises, vapor partial pressure is also increased. The concentration of carbon dioxide in air is very small and relatively constant; therefore the absorbance by carbon dioxide is considered negligible.

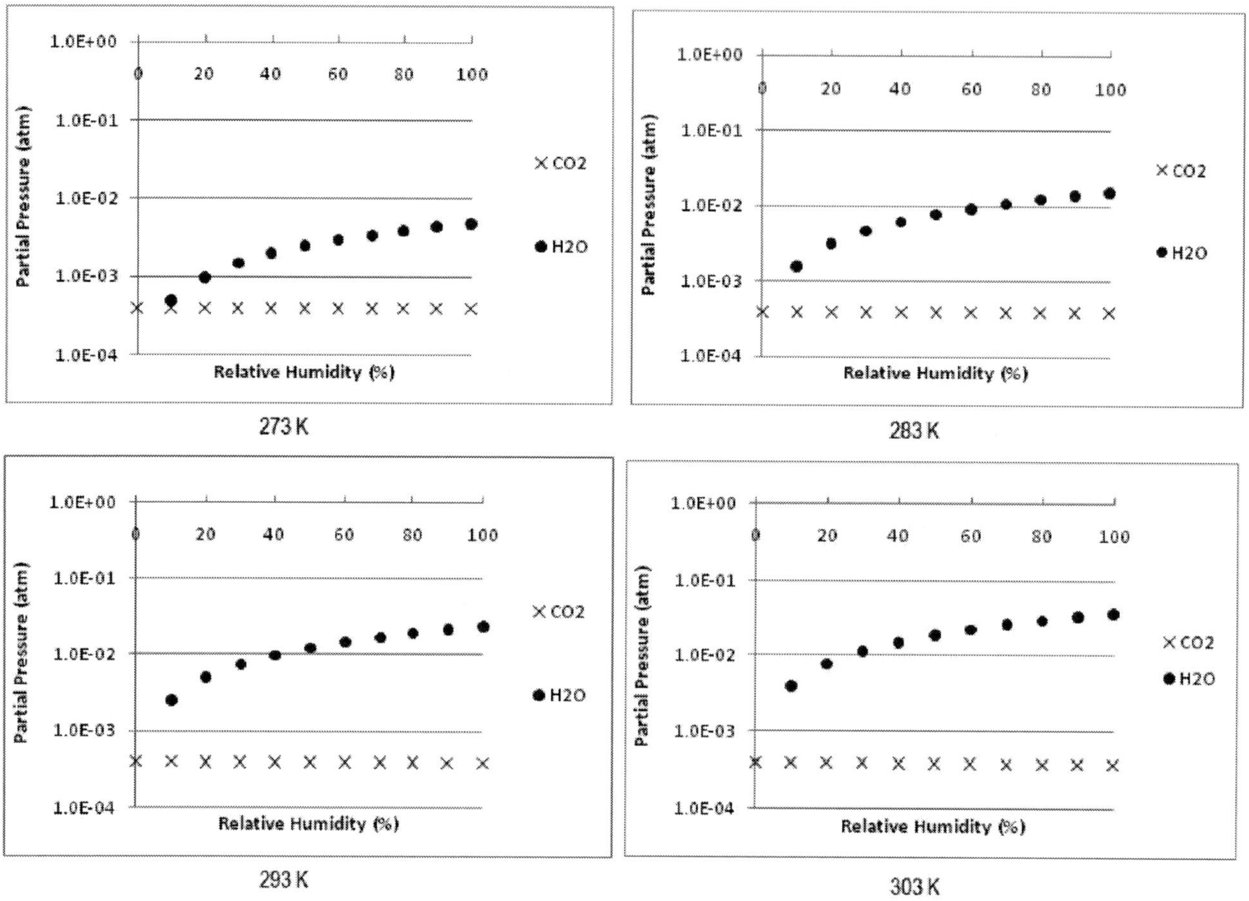

Figure 6. Partial pressures of atmospheric gases as a function of relative humidity at various temperatures

Ambient temperature also gives significant impact on the gas emissivity. Higher ambient temperature also causes an increase in the emissivity weighting factor. Emissivity value corresponds to the weighting factors and absorbing gas partial pressures.

When there is a release of methane gas into the atmosphere, it will be mixed with air and dispersed due to the air turbulence. Therefore, the radiation impinging upon the camera's detector appears not only from the methane gas but also from other gases in the atmosphere including water vapor and carbon dioxide.

The emissivity of the gas mixture is determined using the weighted sum of grey gas model. The graph shown in Figure 7 below shows the emissivity of the gas mixture at various methane gas concentrations where the reduction in gas emissivity at different humidity level can be observed. The graph demonstrates that at the absence of methane gas, higher humidity results in higher emissivity. The emissivity of the gas mixture is reduced at higher humidity.

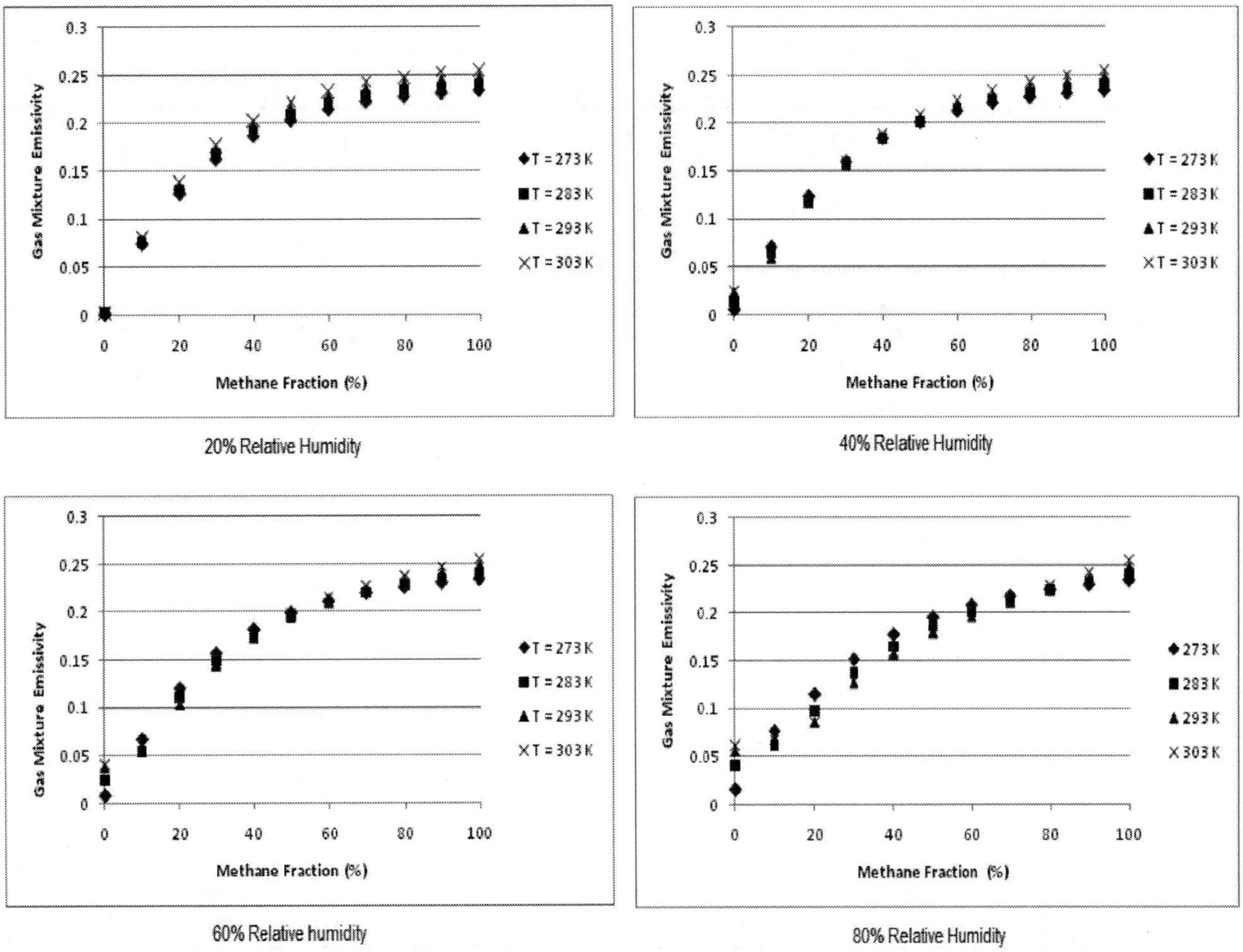

Figure 7. Emissivity of the mixture of dispersed methane gas and atmospheric at various humidity levels and ambient temperatures

Conclusions

Optical imaging using infrared technique has been used in the industry. Although this techniques offer some advantages, they still possess several uncertainties in its use mainly related to the unknown value of target emissivity and radiation measured by the camera detector. The detector is aimed to only visualize hydrocarbon gases; however, the presence of other gases or aerosols in the atmosphere may attenuate the transmittance of infrared radiation from the target gas to the detector. In this paper, effect of environmental conditions i.e. relative humidity and ambient temperature to the performance of infrared optical imaging for methane gas detection and visualization has been studied.

The maximum reduction due to the presence of water vapor in the atmosphere was approximately 25% at methane concentration between 40 – 60 %. At the same humidity level, the concentration of water vapor is higher at higher ambient temperature. Since the water vapor concentration is increasing, the absorption of infrared radiation by water becomes higher and therefore the attenuation in the gas mixture emissivity value becomes more noteworthy.

Si:WO_3 Sensors for Highly Selective Detection of Acetone for Easy Diagnosis of Diabetes by Breath Analysis

Marco Righettoni, Antonio Tricoli and Sotiris E. Pratsinis
Particle Technology Laboratory, Department of Mechanical and Process Engineering, ETH Zurich, Zurich, Switzerland

Introduction

Non-invasive detection of illnesses by human breath analysis[1-2] is an emerging field of medical diagnostics representing a rapid, economic and simple alternative to standard blood analysis[3] and endoscopy[4-5]. The bulk matrix of the breath is a mixture of nitrogen, oxygen, carbon dioxide, water vapor and inert gases. The remaining small fraction consists of more than 1000 volatile traces with concentrations ranging from parts per trillion (ppt) to parts per million (ppm)[1,6] that are either generated in the body (endogenous) or absorbed as contaminants from the environment (exogenous).[7] Some endogenous compounds, including inorganic gases (e.g. NO, CO) and volatile organic compounds (VOCs; e.g. ethane, pentane, ammonia, acetone, ethanol), can be assigned to specific pathologies and thus are utilized as breath markers.[1] In particular, acetone is a selective breath marker to type-1 diabetes.[1,8] It is produced by hepatocytes via decarboxylation of excess Acetyl-CoA.[2] The acetone concentration in the breath increases from 300 - 900 ppb for healthy humans[9] to more than 1800 ppb for diabetic patients.[10] The detection of this small concentration difference (900 ppb) between healthy and ill people in such a complex gas mixture as the human breath requires a highly sensitive and selective acetone sensor.

Several methods have been utilized to analyze trace compounds in the human breath. The most common is gas chromatography (GC) with flame ionization detection (FID),[11] mass spectrometry (MS)[12] or ion mobility spectrometry (IMS).[13] As water vapor, however, can damage the GC column, pre-treatment of the breath is necessary before analysis.[14] Selected ion flow tube mass spectrometry (SIFT-MS)[15] has shown great potential for real-time concentration monitoring of several breath markers (e.g. acetone, ethanol, ammonia and isoprene). A major advantage of the latter is its high selectivity, sufficient sensitivity and low limit of detection (LOD). However, its high cost and limited portability still hinder its application as standard diagnostic tool. Recently, chemo-resistive detectors based on semiconductor metal oxide films have been applied to the analysis of several breath markers.[16-18] These simple devices have low fabrication cost,[19] offer high miniaturization potential,[20] sensitivity[20] and sufficient LOD (ppb concentrations).[19] Among the large number of sensing metal oxides,[19] WO_3[21-22] and in particular its ε-phase is promising for selective and quantitative detection of acetone in ppb concentrations.[18] This is attributed to the spontaneous electric dipole moment of the ε-phase that increases the interaction with analytes having high dipole moment (e.g. acetone).[23] Nevertheless, such WO_3 detectors, have been tested only in ideal conditions (dry air) without accounting for the effect of water vapor which is a major component of the human breath.[24] Furthermore, it is well known that water vapor interferes with the sensing mechanism of semiconductor metal oxides (e.g. SnO_2) decreasing their sensitivity (e.g. to EtOH)[25] and leading to an unreliable response.

Flame aerosol processes, which are used for synthesis of simple nanoparticles commodities, such as carbon black, fumed silica, pigments and catalysis, also offer an efficient and flexible route for synthesis of mixed oxides.[26] The rapid heating and cooling during combustion creates materials (particles or films) with special structure, morphology, composition and high purity resulting in unique products. In addition to the already mentioned FSP-made powder applications, gas sensors are used in environmental (air pollution)-monitoring, process safety control, homeland security, and more recently, such sensors are being explored for early stage detection of diseases from human breath analysis.[18] In particular, SnO_2 nanoparticles for sensing CO, NO_x, and other gases have been made by FSP.[27-28] Recently, FSP has been used also for direct deposition of SnO_2 films onto sensor substrates eliminating the multiple and complex steps of sensor film deposition by wet chemistry.[29] Furthermore, the mechanical stability of these films has been improved by rapid in-situ annealing[20] allowing the integration of flame-based aerosol methods into the industrial fabrication of micro devices, which cannot tolerate temperatures above 400 °C.

Experimental

A flame spray pyrolysis (FSP) reactor was used in combination with a water-cooled substrate holder[29] for synthesis and direct deposition of pure and Si-doped WO_3 nanoparticles films onto Al_2O_3 substrates featuring a set of interdigitated Au electrodes and a sensing area of 1 cm^2. Sensing nanoparticles were prepared as follows: Ammonium (meta) tungstate hydrate and hexamethyldisiloxane were mixed, as dictated by the final Si molar content (0 - 40 mol%), and diluted in a 1:1 (volume ratio) mixture of diethylene glycol monobutyl ether and ethanol with a total metal atom (Si and W) concentration of 0.2 M. This solution was supplied at a rate of 5 ml/min through the FSP nozzle and dispersed to a fine spray with 5 l/min oxygen (pressure drop 1.5 bar). That spray was ignited by a supporting ring-shaped premixed methane/oxygen flame (CH_4 = 1.25 l/min, O_2 = 3.2 l/min). The nozzle-substrate (NS) distance was 20 cm. The films were annealed and mechanically stabilized in-situ by lowering the substrate holder to NS = 14 cm and impinging a particle-free (no metal precursor), xylene-fed spray flame (12 ml/min) onto the film for 30 s.[20] The supporting flame and sheath oxygen flows were those used during FSP deposition.

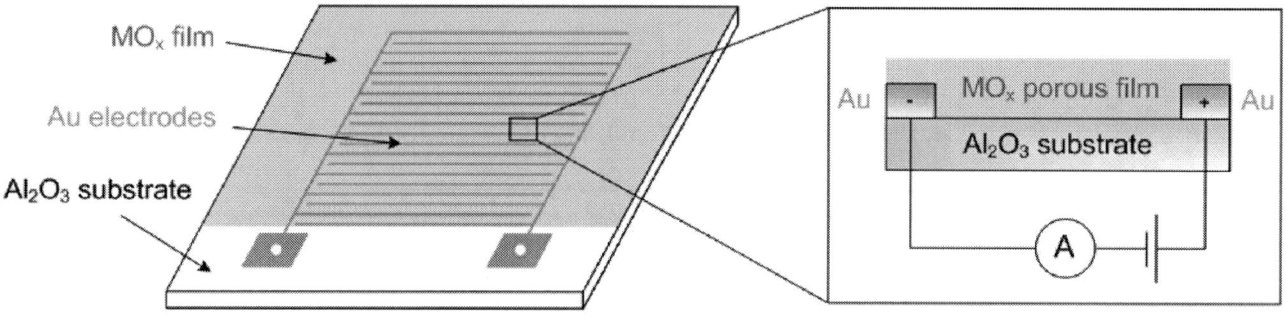

Figure 1 Sensor schematic: a metal oxide (MOx) porous film, in this case pure or Si-doped WO_3, is deposited onto a sensor substrate consisting of an Al_2O_3 support with interdigitated Au electrodes. The analyte concentration is determined by measurement of the nanoparticle film resistance.

Sensor measurements were performed as described in detail elsewhere.[30] Prior to sensing tests, the sensors were kept in an oven at 500 °C for 5 hours to thermally stabilize them and avoid nanoparticle sintering and thereby drift of the sensor signal during testing. Humidified air was generated by bubbling synthetic air through distilled water maintained below room temperature to avoid condensation in the pipes. Acetone or ethanol were controlled by a separate mass flow controller and diluted further with synthetic air to reach the desired concentration. The operating temperature was varied between 325 and 500 °C and measured with a n-type thermocouple placed above the sensor.

Results and Discussion

Here, flame spray pyrolysis has been utilized to synthesize and direct deposit WO_3 nanoparticles onto sensor substrate. The ε-WO_3 phase content was enhanced and thermally stabilized by co-synthesis of SiO_2. Furthermore, the crystal and grain size was tailored by the Si-content. Increasing the Si-content from 0 to 40 mol % reduced the WO_3 crystal (Figure 2, triangles) and grain sizes from 15.2 and 13 to 6.3 and 9.6 nm, respectively. This corresponds to an increase in specific surface area (SSA) from 64 to 117 m^2/g as confirmed also by TEM and ED analysis (Figure 2a-f). At 20 mol% Si, the particle size (Figure 2d,e) is smaller than that of pure WO_3 (a,b). Despite this high Si content, the ED analysis (Figure 2f) confirms the presence of smaller crystallites than those of pure WO_3 (c, bright spots). In fact, the crystal plane of pure WO_3 (Figure 2b) extends continuously and regularly through several particles, whereas at 20 mol% Si content, smaller WO_3 crystallites confined by amorphous SiO_2 regions (Figure 2e) are visible. This should result in increased long term stability and thus avoid baseline drift. Silica doping increased the ε-phase content (Figure 2, circles) from 72 wt% for pure WO_3 to 100 wt% at 10 mol% Si. This is attributed to formation of an interstitial Si-WO_3 solid solution. The resulting nanocomposite optimally combines the high selectivity of the ε-phase of WO_3 and the high sensitivity of small grains.[31] In this way, the sensitivity and thermal stability were increased.

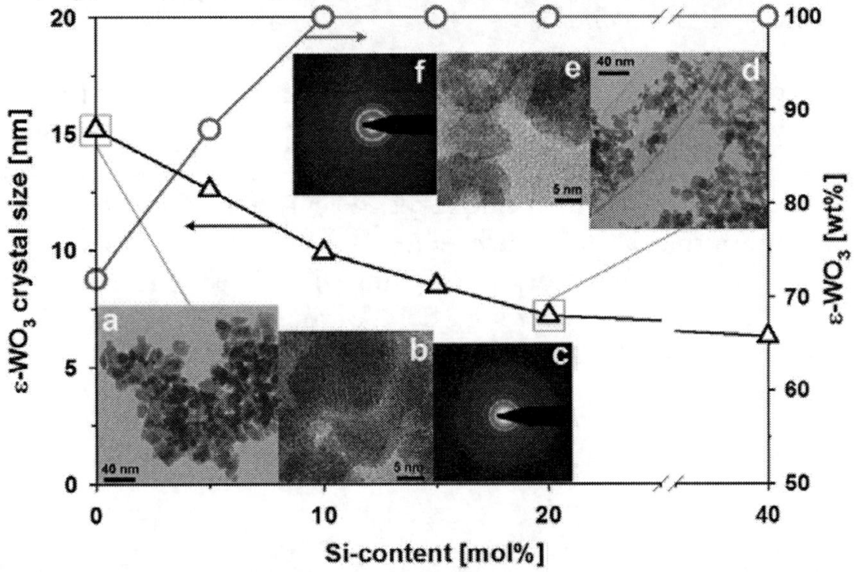

Figure 2 The ε-WO_3 crystal site (triangles) and phase content (circles) as a function of Si-doping of flame-made WO_3-based nanoparticles along with TEM images and electron diffraction (ED) patterns of pure (a-c) and 20 mol% Si-doped (d-f) WO_3 particles. Adapted from [31].

The optimal operating parameters for the detection of acetone have been determined.[30] Chemo-resistive gas-sensors based on tungsten oxide nanoparticles showed high sensitivity to acetone down to 20 ppb in ideal (dry air) and in real condition (90% RH). Figure 3 shows the resistance of a Si-doped WO_3 sensor at realistic conditions and with 20, 50 and 80 ppb acetone at 400 °C.[30] This extremely lower limit of detection (LOD) is comparable to that obtained by more complex systems as SIFT-MS.[15] The high signal-to noise ratio (between 12 and 60) allows precise detection of this extremely low acetone concentration with a relatively simple solid state detector that could be rapidly integrated in portable micro gas sensor systems.[20]

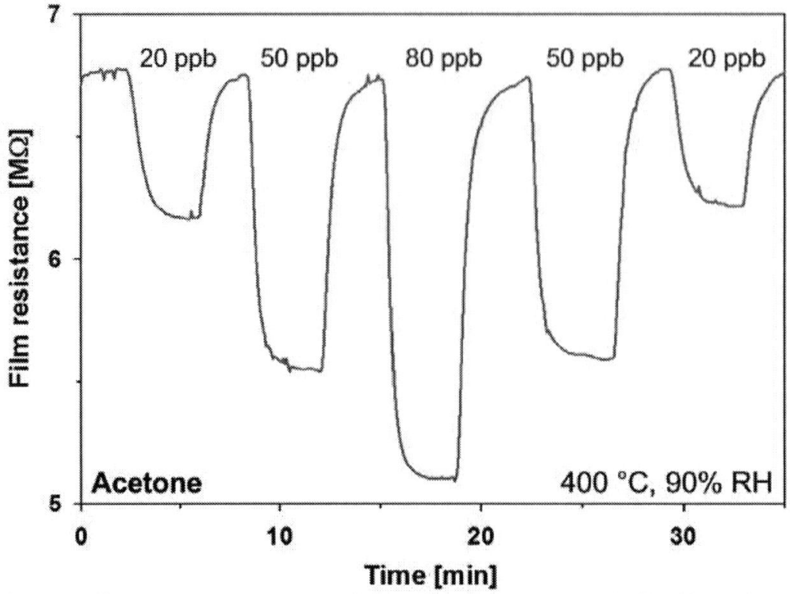

Figure 3 WO_3 sensor resistance (10 mol% Si-doping) at 90% RH and upon exposure to different ultra low concentrations of acetone at 400 °C. Adapted from [30].

The 10 mol% Si-content sensor response to acetone in simulated breath (Figure 4) was below 2 for healthy humans (<0.9 ppm) and above 3.25 for diabetics (>1.8 ppm). This 40% increase in sensor response, even at this high RH, may allow reliable diagnosis of diabetic patients by breath acetone monitoring with the presented portable metal oxide detector. Furthermore, the cross-sensitivity to ethanol and water vapors during acetone detection by ε-WO_3 based gas-sensors was low, creating unique opportunities for using such sensors for non-invasive detection of illnesses (e.g. diabetes) by human breath analysis, which is an emerging field of medical diagnostics, representing a rapid, economic and simple alternative to standard blood analysis and endoscopy.

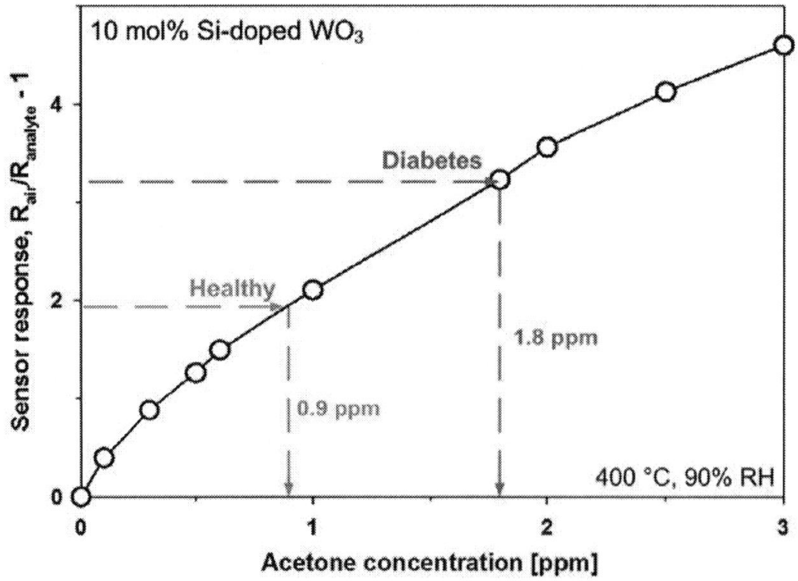

Figure 4 Sensor response for acetone detection of a 10 mol % Si-doped WO_3 sensor with 90% RH at 400 °C. Diabetic patients can be clearly distinguished from healthy humans at 90% RH and upon exposure to different ultra low concentrations of acetone at 400 °C. Adapted from [30].

Conclusions

Pure and Si-doped WO_3 nanoparticles have been synthesized, directly deposited and in situ stabilized onto sensor substrates by FSP. Doping with Si increased the acetone-selective ε-WO_3 phase from 72 to 100 wt%. Those chemo-resistive detectors were utilized for quantitative analysis of acetone concentration, an important breath marker, in ideal (dry air) and realistic (90% RH) conditions. The acetone sensor response at 400 °C in dry air was compared to that of water vapor and ethanol, demonstrating high acetone selectivity for application in medical diagnostics. The sensor responses were maximal below 500 °C while the cross sensitivity to water vapor decreased with increasing operating temperature. Small variations of water vapor content around 90% RH had only minimal effect on the sensor response. Although the sensor response to acetone decreased with increasing RH, the sensor selectivity to acetone was increased. Finally, rapid detection of 20 ppb acetone at realistic (90 % RH) conditions was demonstrated by these low cost, solid state devices based on Si-doped WO_3 nanoparticles, showing great potential for application in noninvasive medical diagnosis.

References

(1) Cao WQ, Duan YX. Breath analysis: Potential for clinical diagnosis and exposure assessment. *Clin. Chem.* 2006;*52*:800-811.

(2) Miekisch W, Schubert JK, Noeldge-Schomburg GFE. Diagnostic potential of breath analysis - focus on volatile organic compounds. *Clin. Chim. Acta* 2004;*347*:25-39.

(3) Lebovitz HE. Diabetic-Ketoacidosis. *Lancet* 1995;*345*:767-772.

(4) Lassen AT, Pedersen FM, Bytzer P, de Muckadell OBS. Helicobacter pylori test-and-eradicate versus prompt endoscopy for management of dyspeptic patients: a randomised trial. *Lancet* 2000;*356*:455-460.

(5) Uemura N, Okamoto S, Yamamoto S, Matsumura N, Yamaguchi S, Yamakido M, Taniyama K, Sasaki N, Schlemper RJ. Helicobacter pylori infection and the development of gastric cancer. *N. Engl. J. Med.* 2001;*345*:784-789.

(6) Phillips M, Herrera J, Krishnan S, Zain M, Greenberg J, Cataneo RN. Variation in volatile organic compounds in the breath of normal humans. *J. Chromatogr. B* 1999;*729*:75-88.

(7) Pleil JD, Lindstrom AB. Exhaled human breath measurement method for assessing exposure to halogenated volatile organic compounds. *Clin. Chem.* 1997;*43*:723-730.

(8) Henderson MJ, Karger BA, Wrenshall GA. Acetone in the Breath - a Study of Acetone Exhalation in Diabetic and Nondiabetic Human Subjects. *Diabetes* 1952;*1*:188-193.

(9) Diskin AM, Spanel P, Smith D. Time variation of ammonia, acetone, isoprene and ethanol in breath: a quantitative SIFT-MS study over 30 days. *Physiol. Meas.* 2003;*24*:107-119.

(10) Deng CH, Zhang J, Yu XF, Zhang W, Zhang XM. Determination of acetone in human breath by gas chromatography-mass spectrometry and solid-phase microextraction with on-fiber derivatization. *J. Chromatogr. B* 2004;*810*:269-275.

(11) Sanchez JM, Sacks RD. GC analysis of human breath with a series-coupled column ensemble and a multibed sorption trap. *Anal. Chem.* 2003;*75*:2231-2236.

(12) Mukhopadhyay R. Dont waste your breath. *Anal. Chem.* 2004;*76*:273A-276A.

(13) Lord H, Yu YF, Segal A, Pawliszyn J. Breath analysis and monitoring by membrane extraction with sorbent interface. *Anal. Chem.* 2002;*74*:5650-5657.

(14) Cheng WH, Lee WJ. Technology development in breath microanalysis for clinical diagnosis. *J. Lab. Clin. Med.* 1999;*133*:218-228.
(15) Smith D, Spanel P. Selected ion flow tube mass spectrometry (SIFT-MS) for on-line trace gas analysis. *Mass Spectrom. Rev.* 2005;*24*:661-700.
(16) Natale CD, Macagnano A, Martinelli E, Paolesse R, D'Arcangelo G, Roscioni C, Finazzi-Agro A, D'Amico A. Lung cancer identification by the analysis of breath by means of an array of nonselective gas sensors. *Biosens. Bioelectron.* 2003;*18*:1209-1218.
(17) Ryabtsev SV, Shaposhnick AV, Lukin AN, Domashevskaya EP. Application of semiconductor gas sensors for medical diagnostics. *Sens. Actuator B-Chem.* 1999;*59*:26-29.
(18) Wang L, Teleki A, Pratsinis SE, Gouma PI. Ferroelectric WO_3 nanoparticles for acetone selective detection. *Chem. Mater.* 2008;*20*:4794-4796.
(19) Eranna G, Joshi BC, Runthala DP, Gupta RP. Oxide materials for development of integrated gas sensors - A comprehensive review. *Crit. Rev. Solid State Mat. Sci.* 2004;*29*:111-188.
(20) Tricoli A, Graf M, Mayer F, Kuhne S, Hierlemann A, Pratsinis SE. Micropatterning layers by flame aerosol deposition-annealing. *Adv. Mater.* 2008;*20*:3005-3010.
(21) Gouma PI, Kalyanasundaram K. A selective nanosensing probe for nitric oxide. *Appl. Phys. Lett.* 2008;*93*:244102.
(22) Khadayate RS, Sali V, Patil PP. Acetone vapor sensing properties of screen printed WO_3 thick films. *Talanta* 2007;*72*:1077-1081.
(23) Woodward PM, Sleight AW, Vogt T. Ferroelectric tungsten trioxide. *J. Solid State Chem.* 1997;*131*:9-17.
(24) Ferrus L, Guenard H, Vardon G, Varene P. Respiratory Water-Loss. *Respir. Physiol.* 1980;*39*:367-381.
(25) Tricoli A, Righettoni M, Pratsinis SE. Minimal cross-sensitivity to humidity during ethanol detection by SnO_2-TiO_2 solid solutions. *Nanotechnology* 2009;*20*:315502.
(26) Strobel R, Pratsinis SE. Flame aerosol synthesis of smart nanostructured materials. *J. Mater. Chem.* 2007;*17*:4743-4756.
(27) Madler L, Sahm T, Gurlo A, Grunwald J-D, Barsan N, Weimar U, Pratsinis SE. Sensing low concentrations of CO using flame-spray-made Pt/SnO2 nanoparticles. *J. Nanopart. Res.* 2006;*8*:783-796.
(28) Sahm T, Madler L, Gurlo A, Barsan N, Pratsinis SE, Weimar U. Flame spray synthesis of tin dioxide nanoparticles for gas sensing. *Sens. Actuator B-Chem.* 2004;*98*:148-153.
(29) Madler L, Roessler A, Pratsinis SE, Sahm T, Gurlo A, Barsan N, Weimar U. Direct formation of highly porous gas-sensing films by in-situ thermophoretic deposition of flame-made Pt/SnO_2 nanoparticles. *Sens. Actuator B-Chem.* 2006;*114*:283-295.
(30) Righettoni M, Tricoli A, Pratsinis SE. Si:WO_3 Sensors for Highly Selective Detection of Acetone for Easy Diagnosis of Diabetes by Breath Analysis. *Anal. Chem.* 2010;*82*:3581-3587.
(31) Righettoni M, Tricoli A, Pratsinis SE. Thermally-stable, silica doped ε-WO_3 for sensing of acetone in the human breath. *Chem. Mater.* 2010;*22*:3152-3157.

(390e) Modeling of a Catalytic Microsensor for the Selective Detection and Quantification of Ethanol in Multi-Component Hydrocarbon Fuel Mixtures

Joseph E. Gatt, Hari Nair, Rong Zhang and Chelsey D. Baertsch, School of Chemical Engineering, Purdue University, West Lafayette, IN

Micro-hotplate-based catalytic gas sensors are designed and fabricated to selectively quantify ethanol in hydrocarbon fuels. This sensor builds upon the concept of a catalytic sensor by introducing a reactant selective catalyst to uniquely oxidize only the target Volatile organic compound (VOC) while allowing all other non-target VOCs to pass by it without generating a response. The microsensor for ethanol quantification utilizing $Fe_2(MoO_4)_3$ as a sensing catalytic substrate at 453 K in an inert environment (hydrocarbon mixtures included alkanes, alkenes, and aromatics) and in mixtures containing methanol were simulated via COMSOL multiphysics software package paired with Matlab v8. The microsensor is essentially modeled as a two dimensional, thermal-catalytic reactor, with dimensions representative of previously fabricated micro-channel reactors in the field (3000 μ m by 500 μ m by 50 μ m). The system is assumed as a cuboid, divided into a gas phase and the solid catalytic layer. In the scenario of ethanol in an inert environment, complete ethanol conversion ranges were found to be 930 ppm ethanol for a sensor with a catalyst layer of 15 μ m. These simulations show how sensor designs can be tailored to meet specific quantification requirements. Temperature profiles for multiple concentrations of ethanol were investigated specifically at the base of the sensor where the RTDs are fabricated within thermalchemical microsensors. A simple empirical relationship can be used to calculate ethanol concentrations in an inert environment based on a single temperature measurement. However, this approach fails in ethanol with methanol mixtures because both methanol and ethanol react on the catalytic sensor and thus multiple concentrations of each alcohol can give rise to similar temperature profiles. The impact that various concentrations of methanol have on the ethanol quantification capability of the sensor was investigated by conducting a series of simulations were ethanol concentrations were held constant at either 100 ppm or 250 ppm while methanol concentrations varied from 0-1250 ppm. By employing a genetic search algorithm, it was demonstrated that unknown concentrations of ethanol and methanol can be estimated from a resulting temperature profile. Accuracy for a simple sensor layout comprised of 4 evenly spaced RTDs proved to be within 8 ppm of ethanol and 26 ppm methanol for a target concentration of 250 ppm ethanol and 100 ppm methanol and was found to improve upon increasing the number of RTDs or their position along the sensor length. Coupling microsensor simulations with a genetic search algorithm will enable quantification of both ethanol and methanol in gas mixtures containing one or more of these reactive gases over a selective oxidation catalyst and can be used to optimize sensor positions to improve detection precision over different concentration ranges.

AUTHOR INDEX

AUTHOR INDEX